SOME IMPORTANT FACTS ABOUT THIS REVISION GU

- IT IS MEANT TO BE USED THROUGHOUT YOUR COURSE AS WELL AS FOR PRE-EXAM REVISION.
- IT IS WRITTEN SPECIFICALLY FOR YOUR SYLLABUS.
- IT CONTAINS ALL THE INFORMATION YOU NEED TO KNOW ...
- ... PRESENTED IN AN INFORMAL, USER FRIENDLY STYLE.
- TOPICS ARE PRESENTED AS SINGLE OR DOUBLE PAGE SPREADS ...
- ... WHICH ARE THEN DIVIDED INTO EASY TO LEARN SUB SECTIONS.
- NOTES ARE WRITTEN IN SHORT, SNAPPY PHRASES ...
- ... WITH KEY FEATURES ...

... HIGHLIGHTED BY GREY BOXES ...

... WRITTEN IN CAPITALS ...

... OR EMPHASISED BY 'BULLET POINTS'.

- AT THE END OF EACH PAGE THERE IS A KEY POINTS SECTION ...
- ... WITH A FULL SET OF SUMMARY QUESTIONS AT THE END OF EACH MODULE.

THOSE SECTIONS OF THE GUIDE WHICH ARE OUTLINED IN RED ARE THE AREAS OF THE SYLLABUS WHICH WILL NOT BE TESTED ON THE FOUNDATION TIER ...
... AND SHOULD ONLY BE LEARNED BY PUPILS ENTERED FOR THE HIGHER TIER ...

SOME IMPORTANT FACTS ABOUT YOUR EXAMINATION

- You will have TWO PAPERS lasting 2 HOURS EACH if you do the HIGHER TIER.
- You will have TWO PAPERS, lasting 1 HOUR 30 MINUTES EACH, if you do the FOUNDATION TIER.
- Each module is tested by a 20 minute multiple choice test consisting of 18 questions.
- There are 25% of the total marks for each of the terminal papers and 25% for the module tests. The remaining 25% is for coursework.

PAPER 1 — Covers modules 1-5.

PAPER 2 — Covers modules 6-9.

HOW TO USE THIS REVISION GUIDE

- Don't just read! LEARN ACTIVELY!
- Constantly test yourself ... WITHOUT LOOKING AT THE BOOK.
- When you have revised a small sub-section or a diagram, PLACE A BOLD TICK AGAINST IT. Similarly, fill in the "Progress and Revision" chart on the inside back cover.
- Jot down anything which will help YOU to remember - no matter how trivial it may seem.
- These notes are highly refined. Everything you need is here, in a highly organised but user friendly format. Many questions depend only on STRAIGHTFORWARD RECALL OF FACTS, so make sure you LEARN THEM.
- THIS IS YOUR BOOK! Use it throughout your course in the ways suggested and your revision will be both organised and successful.

It is important to remember that there is COMMON MATERIAL relating to ATTAINMENT TARGET 3 (MATERIALS AND THEIR PROPERTIES) which candidates are expected to learn for modules 3, 5, 7 and 9 i.e.

STRUCTURE AND CHANGES
ENERGY SOURCES
BONDING AND MATERIALS
UNIVERSAL CHANGES

THIS APPEARS AT THE END OF VOLUME II OF THE REVISION GUIDE UNDER THE TITLE "COMMON MATERIAL FOR MODULES 3, 5, 7 and 9."

CONTENTS

* Numbers in brackets refer to syllabus reference No.

NOTES

MAINTENANCE OF LIFE

LIFE PROCESSES AND CELLS — MoL 1

LIFE PROCESSES COMMON TO ANIMALS AND PLANTS

There are SEVEN characteristics of life, which naturally are exhibited by both plants and animals:

- **REPRODUCTION** - the production of offspring to continue the species.
- **RESPIRATION** - the production of energy by combining energy rich molecules with oxygen.
- **MOVEMENT** - occurs even in plants, but of course much more slowly.
- **SENSITIVITY** - the ability to respond to the immediate environment.
- **GROWTH** - offspring must naturally be smaller than parents, but gradually grow.
- **EXCRETION** - the removal of waste products from within an organism.
- **NUTRITION** - feeding in animals and photosynthesis in green plants.

REMEMBER THESE PROCESSES BY THE NAME "MRS. GREN"!!

FUNCTIONS OF THE PARTS COMMON TO ANIMAL AND PLANT CELLS

CYTOPLASM - most chemical reactions occur here, controlled by ENZYMES.

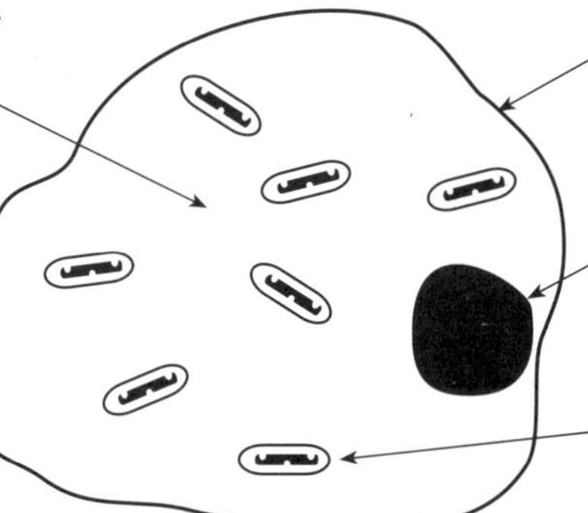

CELL MEMBRANE - Controls movement of substances into and out of the cell.

NUCLEUS - Contains the GENES on CHROMOSOMES which controls the cell's characteristics.

MITOCHONDRIA - Where most energy is released.

LEVELS OF ORGANISATION

- A group of cells with SIMILAR STRUCTURE and FUNCTION is called a TISSUE
- Groups of different tissues form ORGANS
- Organs can work together with other organs to form ORGAN SYSTEMS.
- Several organ systems form an ORGANISM

CELLS ⟶ TISSUES ⟶ ORGANS ⟶ ORGAN SYSTEMS ⟶ ORGANISM

Examples of tissues: MUSCLE, NERVOUS, GLANDULAR.
Examples of organs: HEART, BRAIN, KIDNEY, LUNGS.
Examples of organ systems: CARDIOVASCULAR SYSTEM (Heart, lungs, blood vessels)

KEY POINTS:

- The seven life processes are: Reproduction, Respiration, Movement, Sensitivity, Growth, Excretion and Nutrition.
- Features common to animal and plant cells are: Cytoplasm, Cell Membrane, Nucleus and Mitochondria.
- Cells ⟶ Tissues ⟶ Organs ⟶ Organ Systems ⟶ Organism

THE CIRCULATORY SYSTEM

- Humans have a double circulation i.e. blood passes through the heart TWICE on each full circuit.

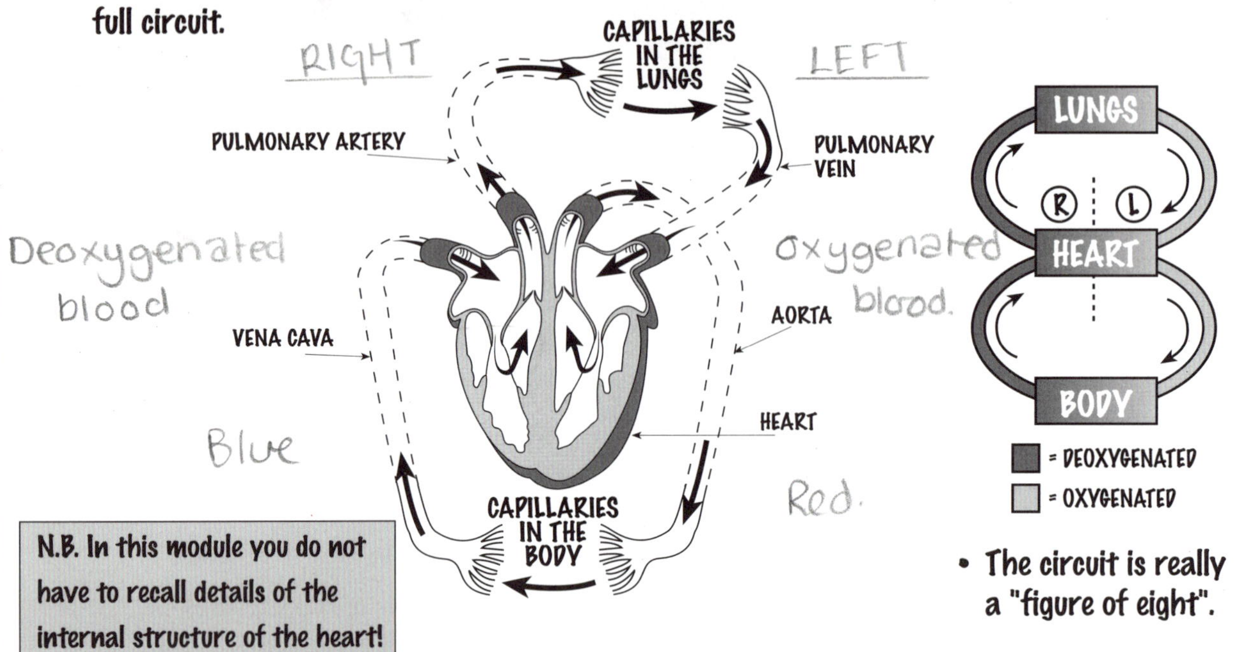

N.B. In this module you do not have to recall details of the internal structure of the heart!

- The circuit is really a "figure of eight".

COMPOSITION AND FUNCTIONS OF THE BLOOD

Blood consists of a watery fluid called PLASMA which carries many other things, the most important being RED BLOOD CELLS, WHITE BLOOD CELLS and PLATELETS.

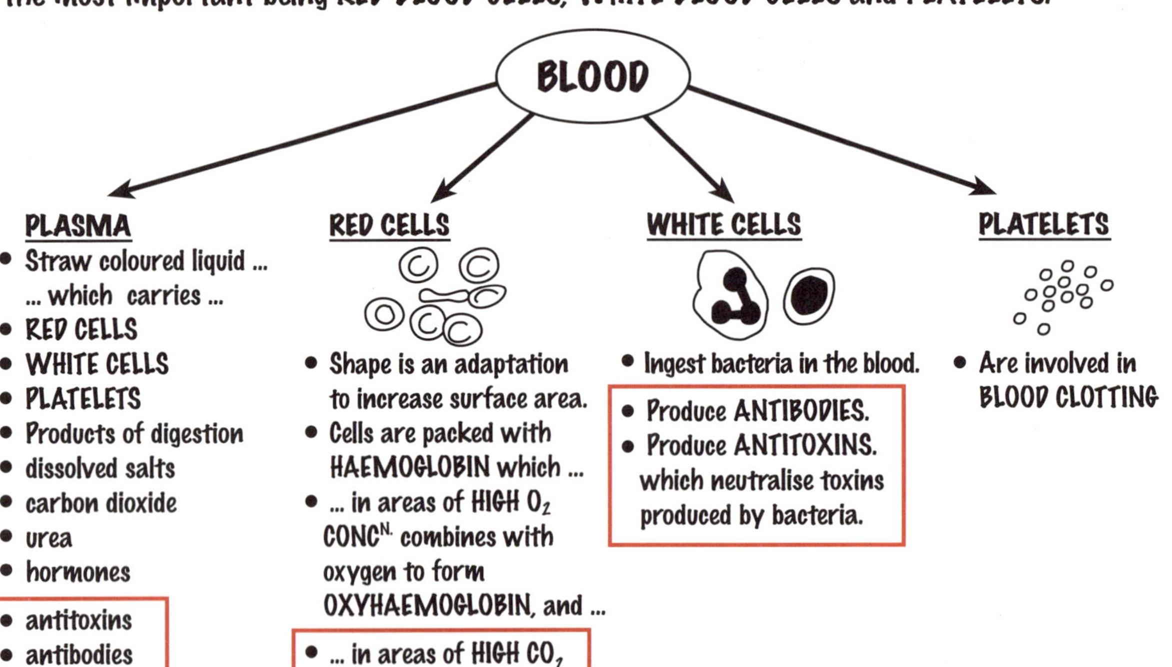

KEY POINTS:

- The Heart acts as a double pump where blood passes through it twice on each full circuit.
- Blood is made up of: Plasma, Red Cells, White Cells and Platelets.

Digestion is the process by which food is broken down into small soluble particles (molecules) by ENZYMES. These can then be absorbed into the blood.

THE DIGESTIVE SYSTEM

TEETH
- To grind and tear food (see MoL.4.)

OESOPHAGUS
- transports food to stomach

STOMACH
- Produces HCl (Hydrochloric acid) ...
- ... to kill bacteria and ...
- ... PROTEASES which ...
- ... digest PROTEINS to AMINO ACIDS.

SALIVARY GLANDS
- Produce SALIVA which contains ...
- ... the enzyme AMYLASE which ...
- ... digests STARCH to SUGAR.

LIVER Stores nuturents
- Stores excess sugar as GLYCOGEN.
- Produces BILE which changes ... ... FATS into DROPLETS.
- Removes POISONS e.g. alcohol from the blood.

PANCREAS
- Produces insulin (see MoL. 11) and LIPASE ...
- ... which digests FATS to FATTY ACIDS.

GALL BLADDER
- Stores bile produced by the liver.

SMALL INTESTINE
- The products of digestion are rapidly absorbed into the blood.

LARGE INTESTINE
- Absorbs water and ...
- ... stores FAECES.

THE VILLUS - For efficient absorption

The Villi in the small intestine ... provide a LARGE, THIN MOIST SURFACE, WITH GOOD BLOOD SUPPLY, for ABSORPTION OF PRODUCTS OF DIGESTION.

PRODUCTS OF DIGESTION AND WATER DIFFUSE THROUGH THE WALLS OF THE VILLI. AND INTO THE BLOOD

A SINGLE VILLUS

GLANDULAR TISSUE

CAPILLARIES

KEY POINTS:

- The digestive system consists of: Teeth, Salivary Glands, Oesophagus, Stomach, Liver, Gall Bladder, Pancreas, Small Intestine and Large Intestine.
- Villi absorb the products of digestion.

ADAPTATION OF TEETH

Mammals have teeth which are specialised to perform specific functions.
In humans these specialisations fall into FOUR broad groups:

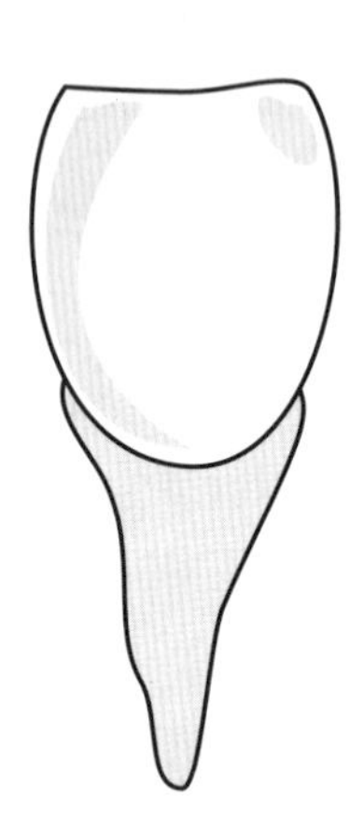
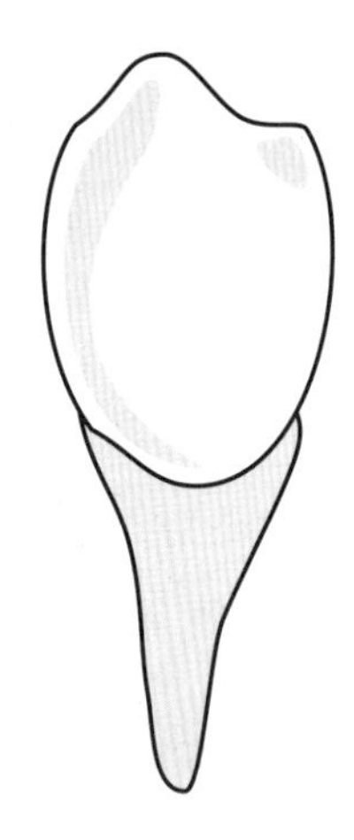
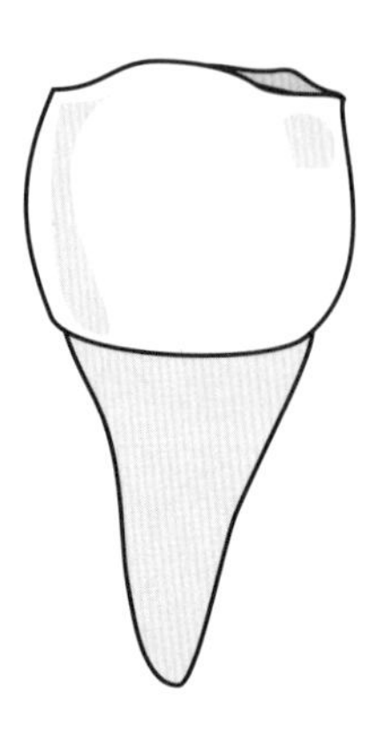
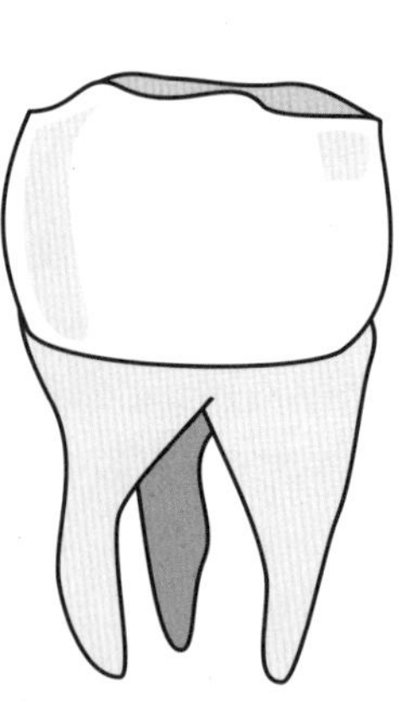

INCISOR

- CHISEL SHAPED ...
- ... for BITING and CUTTING ...
- ... SNIPPING-OFF pieces of meat etc.

CANINE

- POINTED ...
- ... for PIERCING,
- ... TEARING, and ...
- ... HOLDING FOOD.

PREMOLAR and **MOLAR**

- BROAD, "CHUNKY" TEETH,...
- ... with UNEVEN, UPPER SURFACES ...
- ... covered with CUSPS ...
- ... ideal for GRINDING and CRUSHING.

DIGESTION BY ENZYMES - SUMMARY

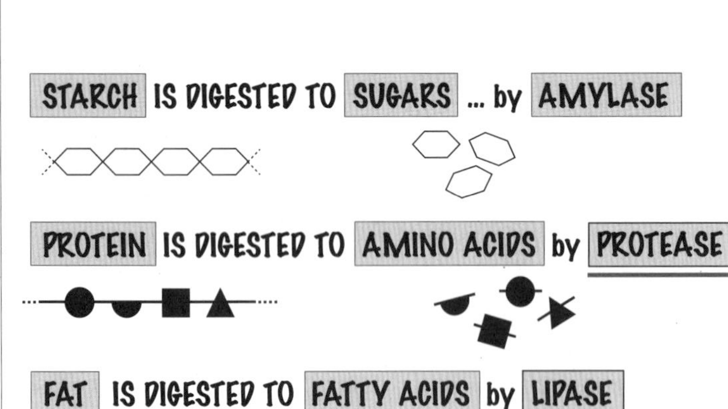

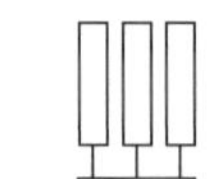
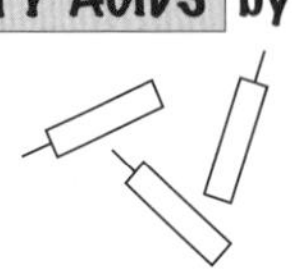

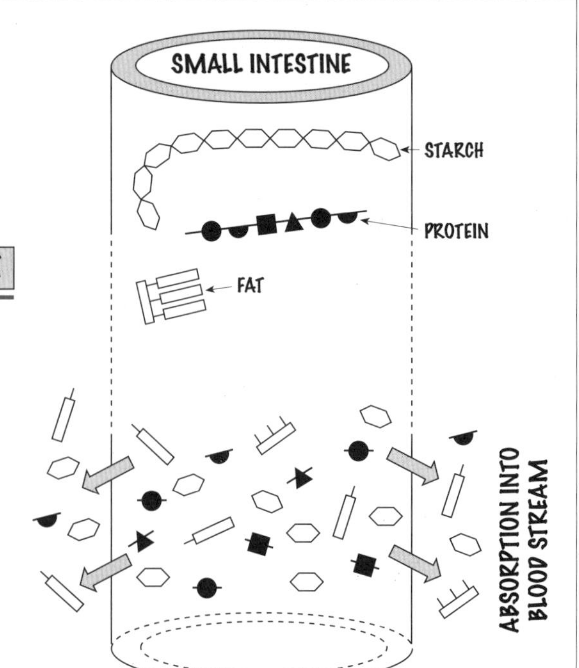

KEY POINTS:

- Incisor, Canine, Premolar and Molar are the four groups of teeth.
- Starch, Protein and Fat are digested by enzymes in the Small Intestine.

NERVOUS SYSTEM I - Organisation

ORGANISATION OF THE NERVOUS SYSTEM

RECEPTORS

- BALANCE (eye and ear)
- SIGHT (eye)
- SMELL (nose)
- TASTE (tongue)
- HEARING (ear)
- TOUCH (skin)
- TEMPERATURE (skin)

BRAIN Coordinates all body activities.

SPINAL CORD Relays "information" to and from the brain to different regions of the body, via specialised nerve cells.

PAIRED SPINAL NERVES Relay impulses from spinal cord to specific areas and vice versa, via specialised nerve cells.

NERVE CELLS - are also called NEURONES and are specialised cells which conduct NERVE IMPULSES. These impulses are ELECTRICAL SIGNALS.

THREE TYPES OF NERVE CELL

EFFECTOR NEURONE
(carries an impulse from the spinal cord to the effector)
e.g. muscle

CONNECTOR NEURONE
(carries an impulse through the spinal cord)

SENSORY NEURONE
(carries an impulse from a receptor to the spinal cord)

OVERALL LAYOUT OF THE SYSTEM

NERVOUS SYSTEM

BRAIN | SPINAL CORD | SENSORY NEURONES | CONNECTOR NEURONES | EFFECTOR NEURONES

CONNECTIONS BETWEEN NEURONES

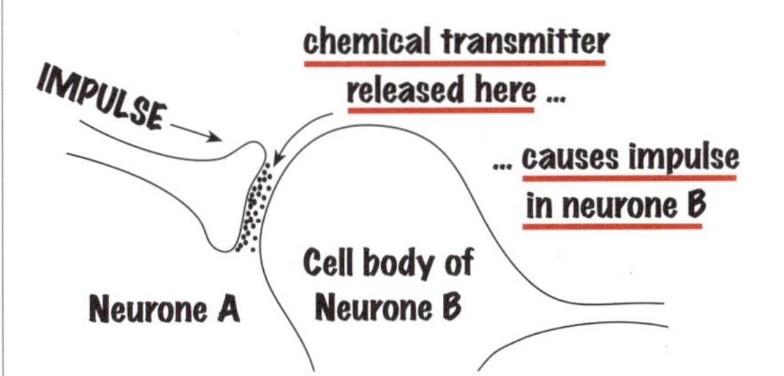

- Neurones do NOT TOUCH EACH OTHER ...
- ... there is a very small gap between them ...
- ... called a **SYNAPSE.**

- When an electrical impulse reaches this gap a CHEMICAL TRANSMITTER ...
- ... is released which causes the electrical IMPULSE to be generated in the next neurone.
- The CHEMICAL TRANSMITTER is then IMMEDIATELY DESTROYED.

KEY POINTS:

- The Nervous System consists of:
  The Brain, Spinal Cord, Sensory Neurones, Connector Neurones and Effector Neurones.
- The gap between neurones is called a Synapse.

NERVOUS SYSTEM II - Reflex Action MoL 6

THE PRINCIPLE OF STIMULUS AND RESPONSE

The pathway for receiving information and then acting upon it is:-

STIMULUS ⇨ RECEPTOR ⇨ SENSORY NEURONE ⇨ COORDINATOR ⇨ EFFECTOR NEURONE ⇨ EFFECTOR ⇨ RESPONSE

	STIMULUS	RECEPTOR	SENSORY NEURONE	COORDINATOR	EFFECTOR NEURONE	EFFECTOR	RESPONSE
e.g.	Insect crawling on skin	Pressure receptor in skin	nerve from skin receptor	Brain via spinal cord	nerve to muscle	muscles in hand	Flick insect away
e.g.	Dust in eye	pain receptor	nerve from pain receptor	Brain	nerve to tear gland	tear glands	tears produced

The EFFECTORS are either MUSCLES or GLANDS.

REFLEX ACTION and the REFLEX ARC

- Reflex action is a rapid automatic response to a stimulus, during which ...
- NERVE IMPULSES are sent by RECEPTORS ...
- ... THROUGH the nervous system to EFFECTORS.

Reflexes are usually DEFENCE MECHANISMS which SPEED UP your RESPONSE TIME by SHORT CIRCUITING THE BRAIN, eg. BLINKING, COUGHING, WITHDRAWAL FROM PAIN.

- the SPINAL CORD acts as the COORDINATOR via ...
- ... a CONNECTOR NEURONE which 'short circuits' the brain ...
- ... by passing IMPULSES from a SENSORY NEURONE directly to an EFFECTOR NEURONE,:-

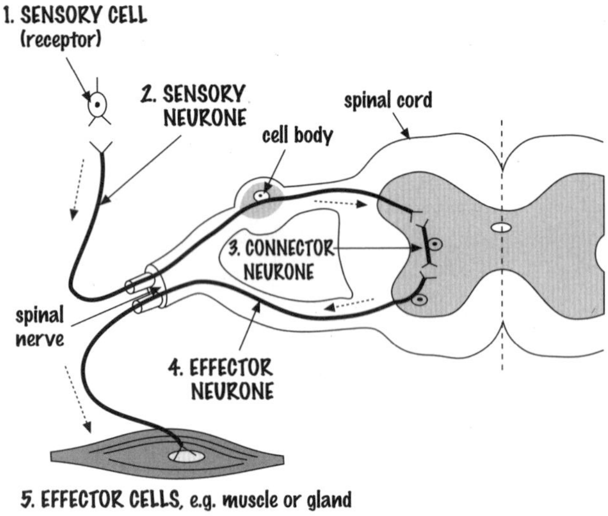

1. SENSORY CELL (receptor)
2. SENSORY NEURONE
cell body
spinal cord
3. CONNECTOR NEURONE
spinal nerve
4. EFFECTOR NEURONE
5. EFFECTOR CELLS, e.g. muscle or gland

STAGES OF THE REFLEX ARC

1. RECEPTORS STIMULATED by PAIN
2. IMPULSES pass along SENSORY NEURONE into SPINAL CORD.
3. Sensory neurone SYNAPSES with CONNECTOR NEURONE, 'short circuiting' the BRAIN.
4. Connector neurone SYNAPSES with EFFECTOR NEURONE, sending IMPULSES down the EFFECTOR NEURONE.
5. These impulses reach MUSCLES causing them to CONTRACT ... bringing about a RESPONSE eg. moving the hand away.

THIS HAPPENS AUTOMATICALLY - WITHOUT CONSCIOUS THOUGHT.

KEY POINTS:

- Receiving information and then acting on it involves the following pathway: Stimulus, Receptor, Sensory Neurone, Coordinator, Effector Neurone, Effector and finally Response.
- A reflex action is a rapid automatic response to a stimulus.

THE EYE

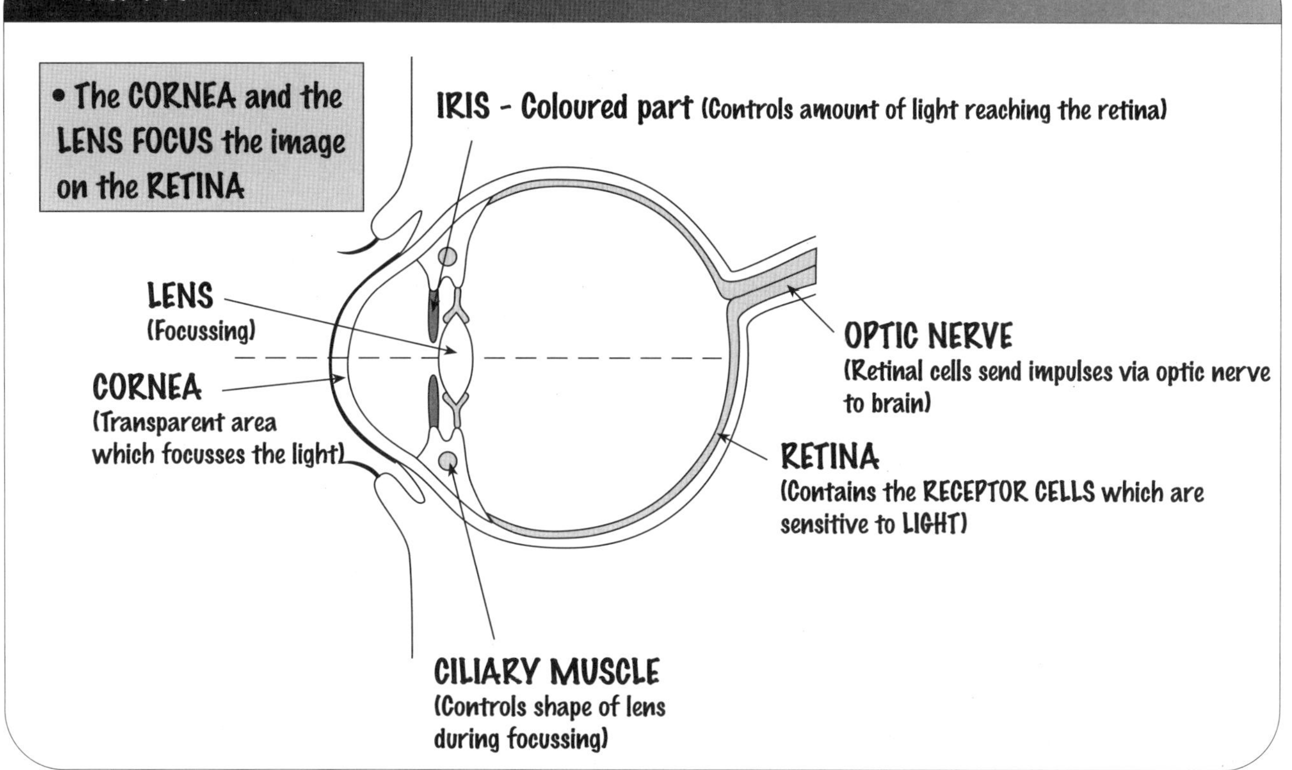

THE EYE AND LIGHT - Focussing

- Image formation occurs when light is REFRACTED by the CORNEA and LENS ...
- ... and FOCUSSED on the RETINA.

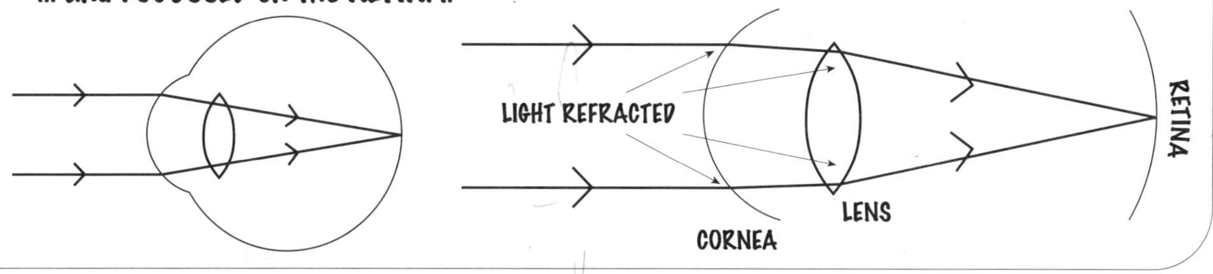

THE EYE AND LIGHT - Controlling the amount of light entering the eye

The IRIS consists of muscle tissue which by contraction of various muscle fibres controls the size of the pupil and therefore the amount of light entering the eye.

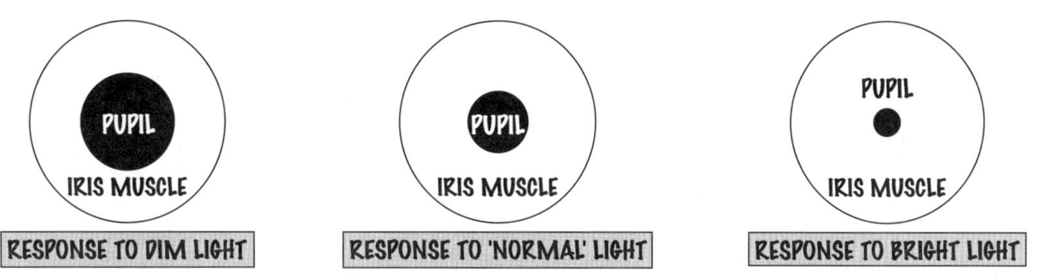

The principle is very similar to the APERTURE CONTROL on a camera.

KEY POINTS:

- The Cornea and the Lens focus the image of an object on the Retina.
- The Iris controls the size of the Pupil.

HORMONES

- Hormones are chemicals secreted by the ENDOCRINE GLANDS ...
- ... directly into the bloodstream in one part of the body ...
- ... and then carried in the blood plasma to a TARGET ORGAN elsewhere.

HORMONAL CONTROL EXAMPLE 1 - INSULIN

- Insulin is produced by ENDOCRINE GLANDS in the PANCREAS in response to high blood sugar...
- ... and is carried in the blood to the LIVER (its TARGET ORGAN).
- The LIVER responds by removing SUGAR from the blood.
- This may occur, for instance, after eating a meal ...
- ... and prevents the BLOOD SUGAR from rising to dangerous levels.

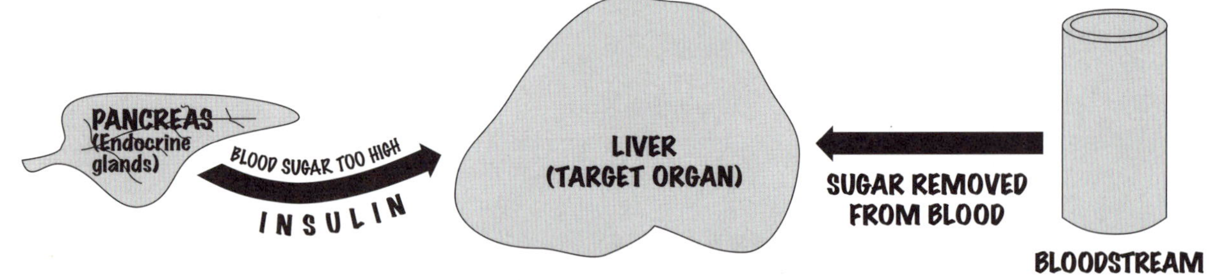

- In the condition called DIABETES, insulin cannot be produced and ...
- ... sufferers may have to control their diet and INJECT INSULIN on a regular basis.

HORMONAL CONTROL EXAMPLE 2 - OESTROGEN AND TESTOSTERONE

- Between the ages of 10 and 16 in girls, and 12 and 17 in boys ...
- ... the sex organs begin to produce the SEX HORMONES which cause ...
- ... the development of the SECONDARY SEXUAL CHARACTERISTICS (PUBERTY).

GIRL	SEX	BOY
OESTROGEN	HORMONE	TESTOSTERONE
OVARIES	ENDOCRINE GLAND	TESTES
• OVULATION AND MENSTRUATION STARTS (i.e. PERIODS). • GROWTH OF BREASTS, UTERUS AND PELVIS. • GROWTH OF PUBIC HAIR, AND ARMPIT HAIR. • DEVELOPMENT OF SOFTER, ROUNDER SHAPE. • FEELINGS OF ATTRACTION TO OPPOSITE SEX.	EFFECT ON TARGET ORGANS. i.e. DEVELOPMENT OF SECONDARY SEXUAL CHARACTERISTICS.	• PRODUCTION OF SPERM. • GROWTH OF MUSCLES AND PENIS. • VOICE BECOMES DEEPER. • GROWTH OF PUBIC HAIR, FACIAL HAIR AND BODY HAIR. • FEELINGS OF ATTRACTION TO OPPOSITE SEX.

All these changes take place relatively slowly unlike a NERVOUS RESPONSE!

- Manufactured hormones may be given to women in order to promote fertility ...
- The type of hormone given is one which STIMULATES EGGS TO MATURE IN THE OVARIES.

KEY POINTS:

- Hormones are chemicals secreted by the Endocrine Glands.
- Insulin controls the Blood Sugar Level.
- Oestrogen and Testosterone control the development of an individual during puberty.

HOMEOSTASIS I - The Excretory System

- This is the **MAINTENANCE OF A CONSTANT INTERNAL ENVIRONMENT.**
- Waste products can't just be left to accumulate inside the body ...
- ... because they would change our **INTERNAL ENVIRONMENT.**
- Similarly, other levels need to be controlled within fairly narrow tolerances.

THE EXCRETORY SYSTEM - Removing waste and controlling water content

DIAPHRAGM
VENA CAVA
AORTA
LEFT KIDNEY
RENAL VEIN
RENAL ARTERY
Blood supply to and from the kidneys
URETER (carries urine from kidney to bladder)
BLADDER (stores urine)
URETHRA (carries urine from bladder to outside)

THE NEPHRON - the functional unit of the kidney

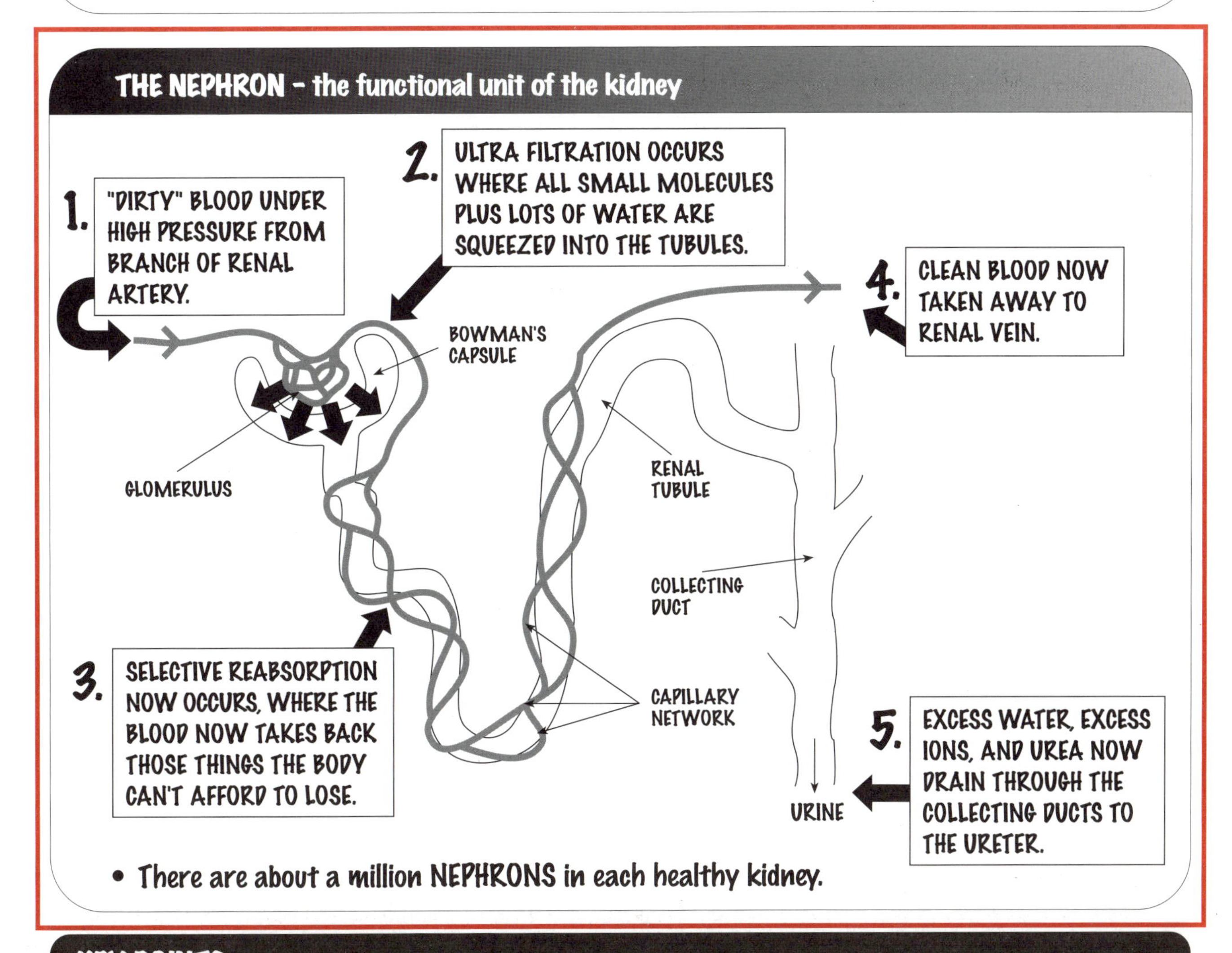

- There are about a million **NEPHRONS** in each healthy kidney.

KEY POINTS:

- Homeostasis is the maintenance of a constant internal environment.
- The Excretory System removes Urea and controls water content.

TREATMENT OF KIDNEY FAILURE

- If one kidney fails then there is no reason why a person shouldn't lead a normal life ...
- However, if both kidneys fail then either DIALYSIS, or a KIDNEY TRANSPLANT is necessary to preserve life.

1. DIALYSIS

- Blood, taken from a vein ...
- ... runs into a DIALYSIS MACHINE, where it comes into close contact ...
- ... with a SELECTIVELY PERMEABLE MEMBRANE, which seperates it ...
- ... from the DIALYSIS FLUID.
- ... WASTE diffuses from the BLOOD into the DIALYSIS FLUID.

BLOOD

DIALYSIS FLUID

- This must be done two or three times per week.

SELECTIVELY PERMEABLE MEMBRANE

2. KIDNEY TRANSPLANT

- This involves taking a kidney from a suitably healthy donor and ...
- ... surgically attaching it to the recipient.
- The donor must be compatible in order to avoid tissue rejection.

	DIALYSIS	KIDNEY TRANSPLANT
ADVANTAGES	No rejection can occur.	No need for regular dialysis - can lead less restricted life.
DISADVANTAGES	Regular sessions in hospital or at home on dialysis machine (10 hours).	Rejection can occur where body's defence system "attacks" the kidney.

CONTROL OF WATER CONTENT - An example of negative feedback (See MoL.11)

- Water content is controlled using the PRINCIPLE of NEGATIVE FEEDBACK, ...
- ... in which ANTI-DIURETIC HORMONE (A.D.H.) is released by the PITUITARY GLAND ...
- ... and acts on the KIDNEY TUBULES.

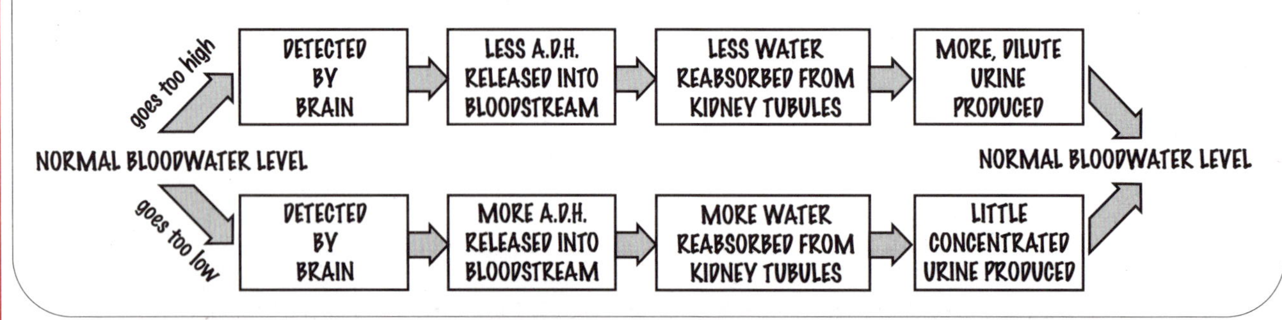

KEY POINTS:

- If Kidney failure occurs then Dialysis or a Kidney Transplant is necessary.
- Water content is controlled by A.D.H. released by the Pituitary Gland.

- Negative feedback (see MoL.10.) simply means that the actual response ...
- ... is responsible for switching itself off (Think of a thermostatically controlled oven).
- NEGATIVE FEEDBACK SYSTEMS are very important in HOMEOSTASIS.

EXAMPLE 1: Control of blood sugar by the hormone insulin

Blood sugar needs to be controlled so that there's always enough in the blood but not too much so the blood doesn't become too thick and syrupy!

When the blood sugar is too high ...

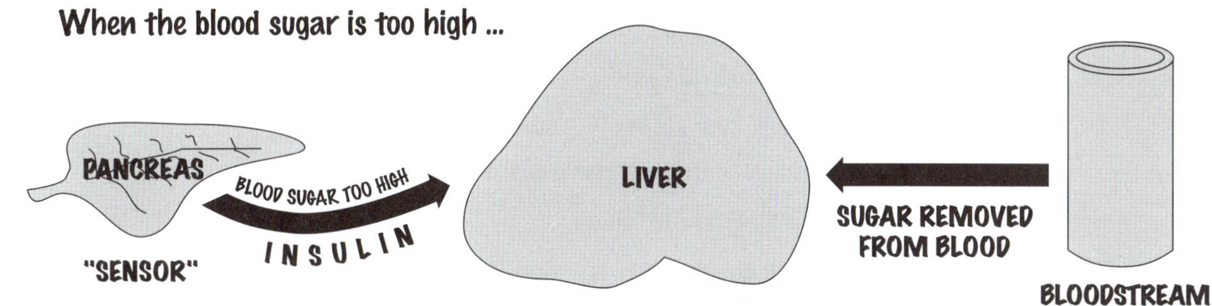

When the blood glucose becomes too low this is automatically detected by the "sensor", the pancreas which stops producing insulin.

EXAMPLE 2: CONTROL OF BODY TEMPERATURE - by the Nervous System

This is controlled by the NERVOUS SYSTEM, and is again a NEGATIVE FEEDBACK SYSTEM.

- The CORE TEMP of the body should be kept at around 37°C (best for enzymes!)
- MONITORING AND CONTROL is done by the THERMO REGULATORY CENTRE in the BRAIN ... ... which monitors the temperature of the blood flowing through it.

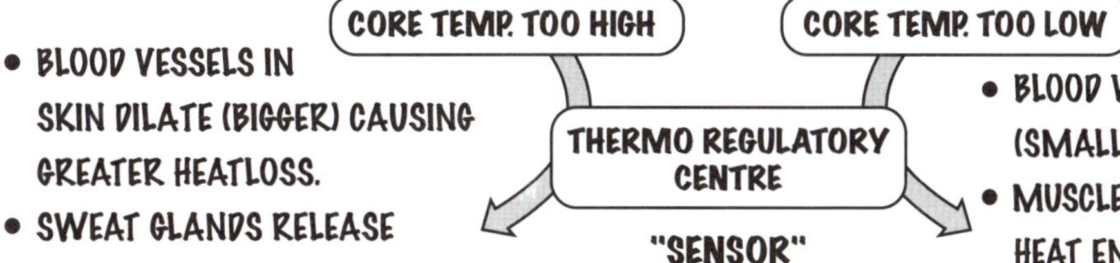

- BLOOD VESSELS IN SKIN DILATE (BIGGER) CAUSING GREATER HEATLOSS.
- SWEAT GLANDS RELEASE SWEAT WHICH EVAPORATES CAUSING COOLING.

- BLOOD VESSELS IN SKIN CONSTRICT (SMALLER) REDUCING HEAT LOSS.
- MUSCLES START TO 'SHIVER' CAUSING HEAT ENERGY TO BE RELEASED VIA RESPIRATION IN CELLS.

THE SKIN'S ROLE IN TEMPERATURE CONTROL

- The skin forms a WATERPROOF and GERM-PROOF protective layer around the body.

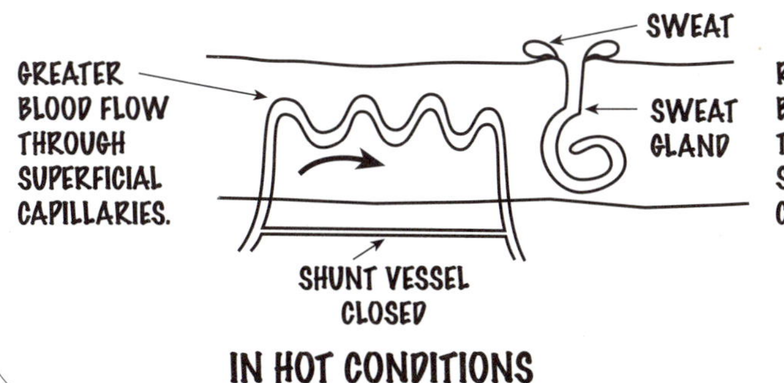

KEY POINTS:

- Blood Sugar is controlled by the hormone Insulin.
- Body Temperature is controlled by the Thermoregulatory Centre in the Brain.

MAINTENANCE OF LIFE SUMMARY QUESTIONS

1. List the seven characteristics of living things.
2. Draw an animal cell and label as many things as you can.
3. What happens at mitochondria?
4. Write down the five levels of organisation.
5. Draw a rough diagram illustrating the "figure of eight" circulation.
6. What are the main vessels entering and leaving the heart?
7. What does plasma contain?
8. What do Red cells, White cells and platelets do?
9. What is the job of antibodies and antitoxins?
10. What are the functions of the oesophagus, stomach and small intestine?
11. What are the functions of the salivary glands, pancreas and large intestine?
12. What are the functions of the liver and gall bladder?
13. Explain which features of the inner surface of the small intestine make it well adapted for efficient absorption.
14. What do Amylase, Protease and Lipase do?
15. What are the four basic teeth types in humans.
16. What are their functions?
17. What does the nervous system consist of?
18. Draw diagrams of the three basic types of nerve cell (neurone).
19. What happens at a synapse?
20. Name the different receptors found in the human body.
21. Write down the pathway leading from a stimulus to a response.
22. What is the purpose of reflex action?
23. Explain the stages in a reflex arc.
24. What are effectors?
25. Write down the six most important parts of the eye and say what each part does.
26. Draw a rough diagram to show how the lens and cornea refract light to focus it onto the retina.
27. Describe the response of the iris to changing light intensity.
28. What are hormones?
29. Describe the action of insulin.
30. Name the hormone involved, and the changes which occur during development of female secondary sexual characteristics.
31. Name the hormone involved, and the changes which occur during development of male secondary sexual characteristics.
32. What is homeostasis?
33. What are the main parts of the excretory system?
34. Describe the 5 stages in the functioning of the nephron.
35. What is meant by Dialysis?
36. Compare the advantages and disadvantages of dialysis with the alternative of having a kidney transplant.
37. Describe how A.D.H. controls the water content of the blood.
38. What is meant by negative feedback?
39. Describe how body temperature is controlled by the nervous system.
40. Describe in detail the skin's role in control of body temperature.

MAINTENANCE OF THE SPECIES

REPRODUCTION AND INHERITANCE I - Meiosis and Mitosis MoS 1

MEIOSIS AND SEXUAL REPRODUCTION

- SEXUAL REPRODUCTION involves the production of MALE and FEMALE GAMETES by MEIOSIS ...
- ... followed by the FUSION OF THESE GAMETES (FERTILISATION), which results in ...
- ... the development of a FOETUS.
- Male gametes (SPERM) are produced in the TESTES. • Female gametes (EGGS) are produced in the OVARIES.

MEIOSIS - Halving chromosome number with TWO DIVISIONS

Let's consider a cell with 4 chromosomes. One LONG pair and a SHORT pair.

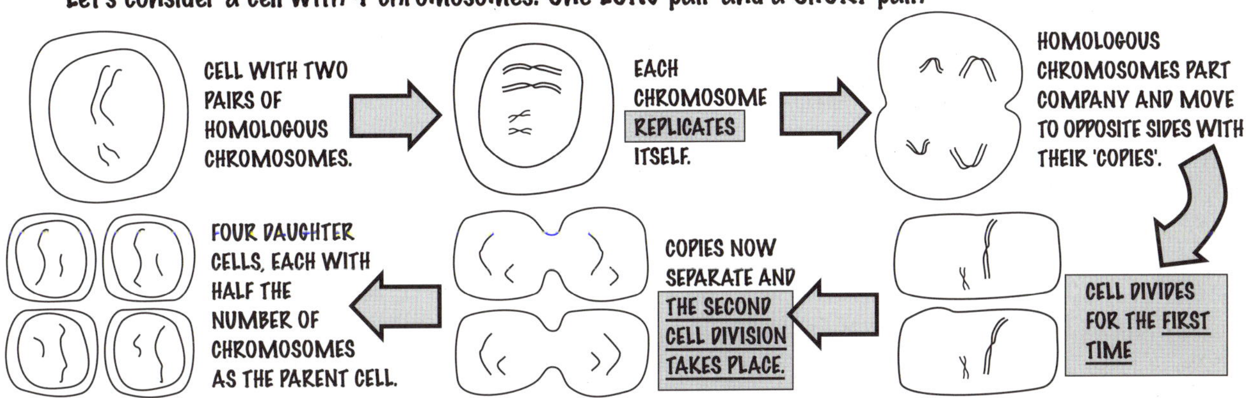

- Both EGGS and SPERM have only half the 'normal' number of chromosomes ...
- ... i.e. 23 NOT 23 PAIRS IN HUMANS.
- Because of the behaviour of the chromosomes, ...
- ... some REASSORTMENT OF GENETIC MATERIAL takes place ... which promotes VARIATION (see MoS.7).

MITOSIS AND ASEXUAL REPRODUCTION

- ASEXUAL REPRODUCTION results in the production of GENETICALLY IDENTICAL OFFSPRING by MITOSIS.
- MITOSIS also occurs in all organisms which reproduce sexually ...
- ... but only in non-reproductive cells for GROWTH and REPAIR.
  (Growth being cell enlargement, cell division and cell specialisation.)

MITOSIS - maintaining chromosome number with ONE DIVISION

Let's consider a cell with just 4 chromosomes. One LONG pair and a SHORT pair.

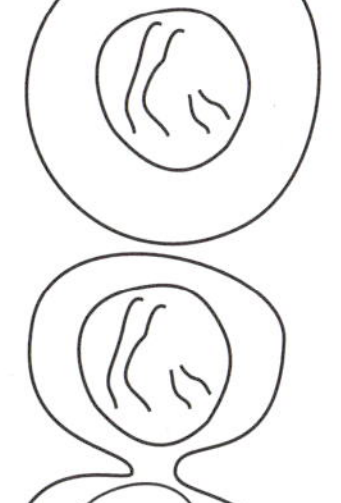

CELL WITH TWO PAIRS OF HOMOLOGOUS CHROMOSOMES.

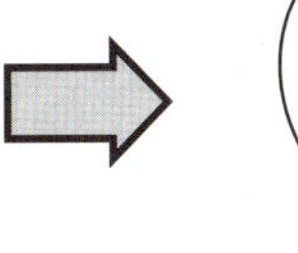

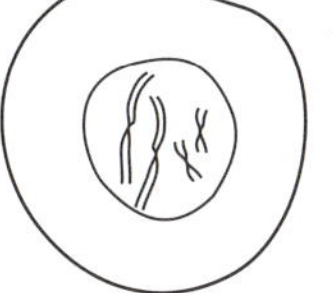

EACH CHROMOSOME REPLICATES ITSELF.

NUCLEAR MEMBRANE DISAPPEARS AND THE 'COPIES' ARE PULLED APART. CELL NOW DIVIDES FOR THE ONLY TIME.

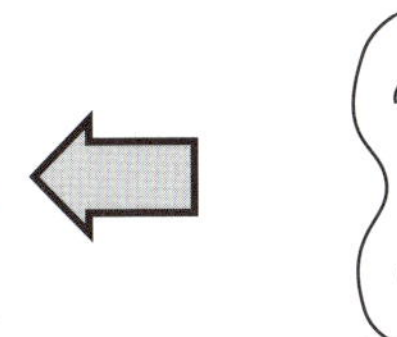

EACH 'DAUGHTER' CELL HAS SAME NUMBER OF CHROMOSOMES AS THE PARENT CELL, AND IS GENETICALLY IDENTICAL

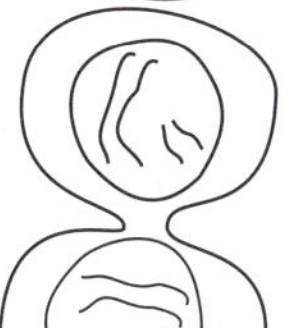

- The production of GENETICALLY IDENTICAL OFFSPRING is known as CLONING.

KEY POINTS:

- Sexual reproduction involves the production of male and female gametes by Meiosis.
- Asexual reproduction involves the production of genetically identical offspring by Mitosis.

CHROMOSOMES, GENES and D.N.A.

- **Genes are the units of inheritance, each one controlling a particular characteristic.**

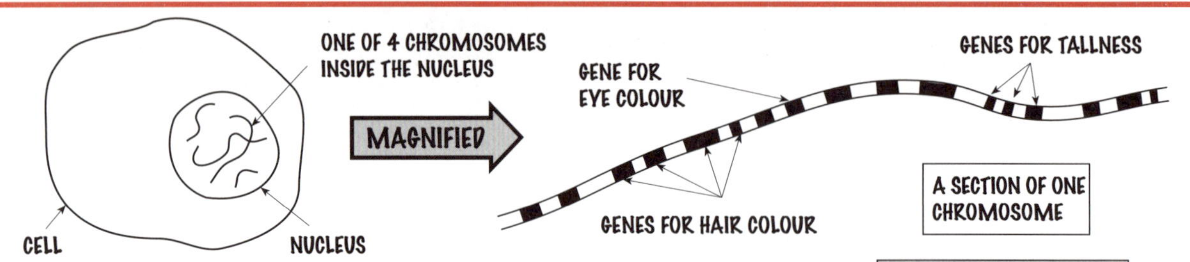

- In NORMAL CELLS (non-reproductive) CHROMOSOMES ARE FOUND IN PAIRS (HOMOLOGOUS PAIRS) ...
- ... 23 PAIRS IN HUMANS. (One of each pair from the MALE and the other from the FEMALE)
- This number differs from species to species.
- Genes are SECTIONS OF D.N.A. which DETERMINE INHERITED CHARACTERISTICS.
- This information is in the form of a CODE ...

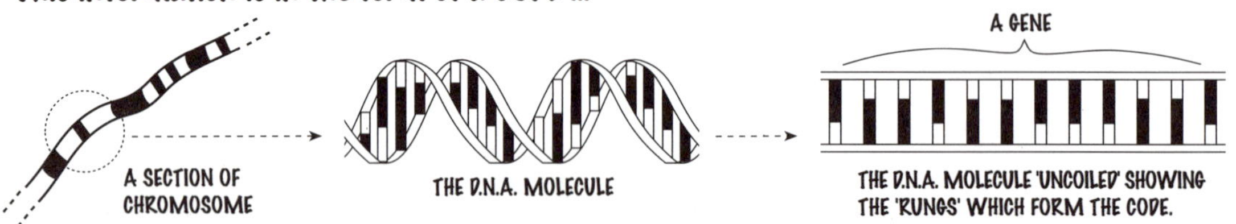

ALLELES - Dominant and recessive

- Since body cells contain PAIRS OF HOMOLOGOUS CHROMOSOMES ...
- ... the GENES which control particular characteristics ALSO COME IN PAIRS ...
- ... which may be in different forms called ALLELES.

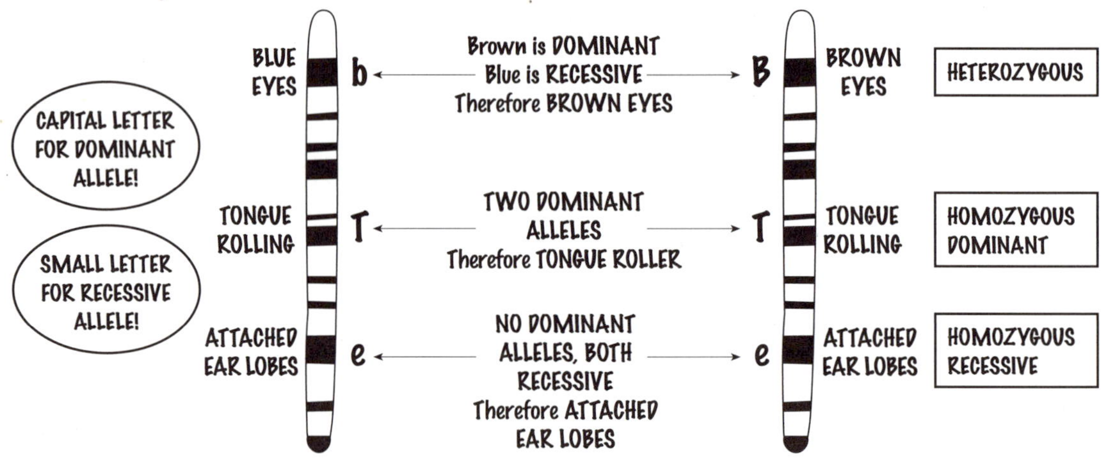

- DOMINANT ALLELES EXPRESS THEMSELVES IF PRESENT ONLY ONCE ...
  ... so an individual can be HOMOZYGOUS DOMINANT (BB) or HETEROZYGOUS (Bb) for brown eyes.
- RECESSIVE ALLELES EXPRESS THEMSELVES ONLY IF PRESENT TWICE ...
  ... so an individual can only be HOMOZYGOUS RECESSIVE (bb) for blue eyes.

<u>TWO IMPORTANT DEFINITIONS:</u>

<u>'GENOTYPE'</u> - This refers to the particular pair of alleles representing a particular characteristic e.g. A person may have the genotypes BB, Bb, or bb for eye colour.

<u>'PHENOTYPE'</u> - This refers to the outward expression of a genotype, e.g. the genotype bb from the example above would have a PHENOTYPE of blue eyes, whereas BB or Bb would give a brown-eyed PHENOTYPE.

KEY POINTS:

- Chromosomes are found in Homologous Pairs.
- Dominant alleles express themselves if present only once.
- Recessive alleles express themselves only if present twice.

GENDER DETERMINATION

- Humans have 23 pairs of HOMOLOGOUS CHROMOSOMES ...
- ... one pair of which are the SEX CHROMOSOMES.
- In females these are IDENTICAL (Homologous) and are called the X chromosomes.
- In males ONE IS MUCH SHORTER THAN THE OTHER and they're called the X and Y chromosomes.

THE POSSIBLE PERMUTATIONS

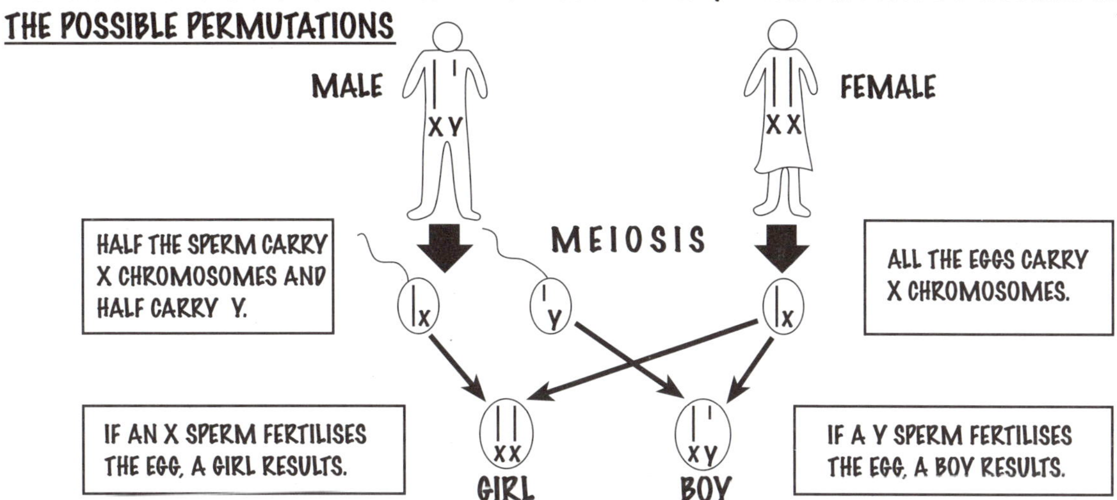

- Like all homologous chromosomes, the SEX CHROMOSOMES SEPARATE DURING MEIOSIS ...
- ... resulting in just one in each sperm or egg.

MONOHYBRID INHERITANCE

- MONOHYBRID INHERITANCE is the inheritance of a feature controlled by ONE PAIR OF ALLELES.
- So, for instance the inheritance of eye colour in an example of this ...
- ... and we'll use this to show the way of tackling GENETIC DIAGRAMS.

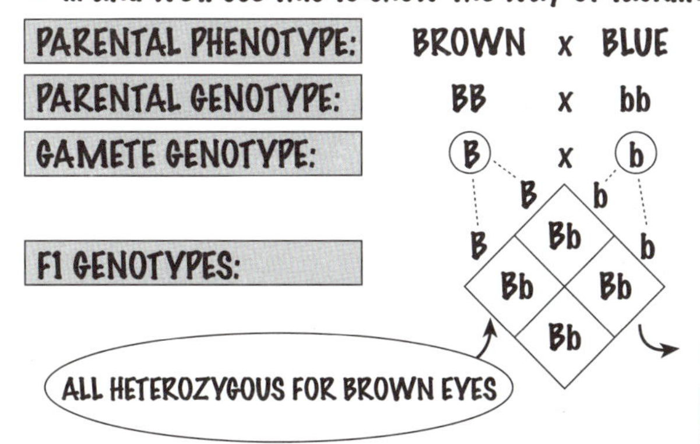

If we take two of these individuals and cross them:-

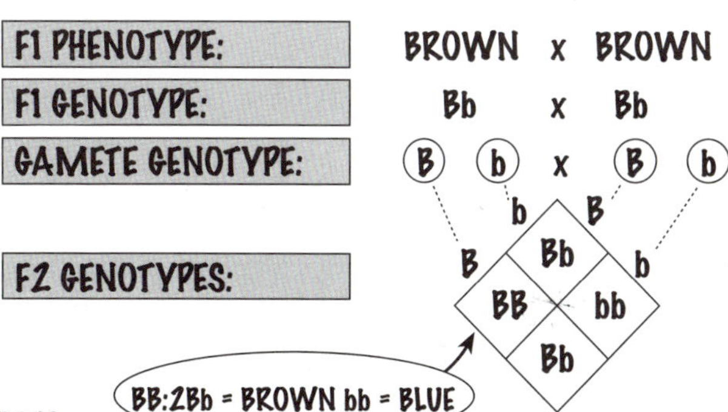

The F1 generation is the first generation produced by the pure breeding parents.
The F2 generation is the first generation produced by the F1's i.e. the "grandchildren."

- There are more examples of these on the next page.

KEY POINTS:

- Males have two different sex chromosomes: X and Y. Females only have X chromosomes.
- Monohybrid Inheritance is the inheritance of a feature controlled by one pair of alleles.

CYSTIC FIBROSIS - caused by recessive alleles

- Cystic Fibrosis can be passed on by parents, neither of whom have the disease ...
- ... if each is carrying just one RECESSIVE allele for the condition.
- It is a disorder of cell membranes causing THICK and STICKY MUCUS ...
- ... especially in the LUNGS, GUT and PANCREAS, which leads to various complications.

F1 PHENOTYPE:	CARRIER	x	CARRIER
F1 GENOTYPE:	Ff	x	Ff
GAMETE GENOTYPE:	(F) (f)	x	(F) (f)

F2 GENOTYPES:

	f	F
F	FF	Ff
f	Ff	ff

FF = normal individual
2Ff = carriers
ff = sufferer from cystic fibrosis

- This particular cross would result in a 1 in 4 chance of producing a sufferer.

SICKLE CELL ANAEMIA - caused by a recessive allele

- Sickle-cell anaemia can be passed on by parents neither of whom has the disease ...
- ... if each is carrying just ONE RECESSIVE ALLELE for the condition.
- Sufferers produce abnormally shaped red blood cells (SICKLE-SHAPED!) ...
- ... and experience general weakness and ANAEMIA.

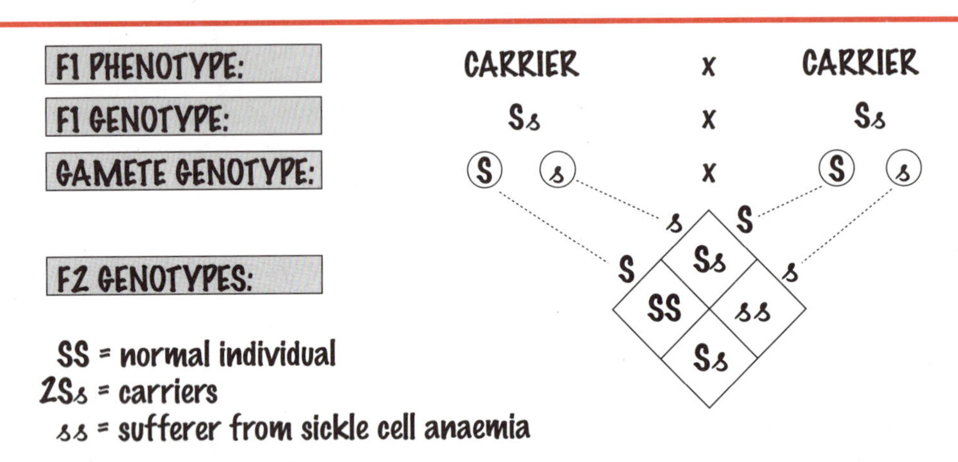

- This cross would also result in a 1 in 4 chance of producing a sufferer.

- The HETEROZYGOUS (Ss) INDIVIDUALS also show up to 50% sickling of cells ...
- ... but show an INCREASED RESISTANCE TO MALARIA which maintains the gene in the population ...
- ... in areas where MALARIA is prevalent.

KEY POINTS:

- Cystic Fibrosis is caused by Recessive Alleles and is a disorder of the cell membranes.
- Sickle Cell Anaemia is caused by Recessive Alleles where sufferers produce abnormally shaped red blood cells.

GENETIC ENGINEERING

- Genetic engineering involves moving sections of D.N.A. from one species ...
- ... to another to manufacture useful Biological products e.g. INSULIN ...

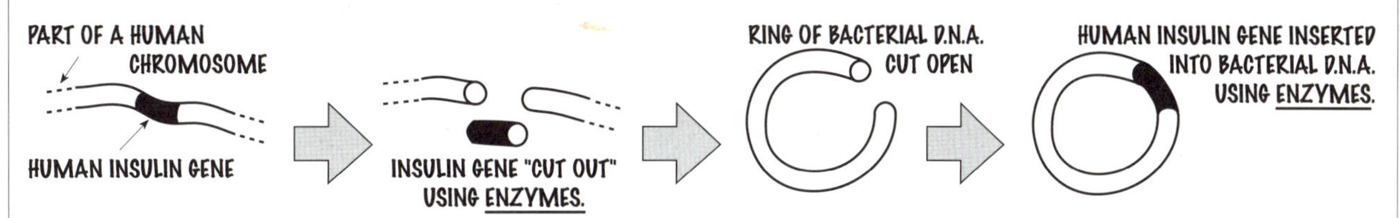

- The bacterium can now carry out the chemistry encoded in the Human Insulin Gene ...
- ... i.e. it can make INSULIN.
- Also, transferred genes are copied exactly every time the bacterium divides, so ...
- ... a culture containing billions of insulin-producing bacteria can be produced (GENE CLONING).
- This makes it possible to produce COMMERCIAL QUANTITIES OF INSULIN, for the treatment of DIABETES.

THE WORK OF MENDEL

- Gregor Mendel was born in Austria in 1822, and was ordained as a priest in 1847.
- His work on PEA PLANTS marks the start of modern genetics.
- He started with PURE-BREEDING Parent plants (i.e. characteristics appear unchanged each generation when self-pollinated).
- ... and produced the F1 generation which he allowed to self pollinate to produce the F2 generation.

Here's a summary:

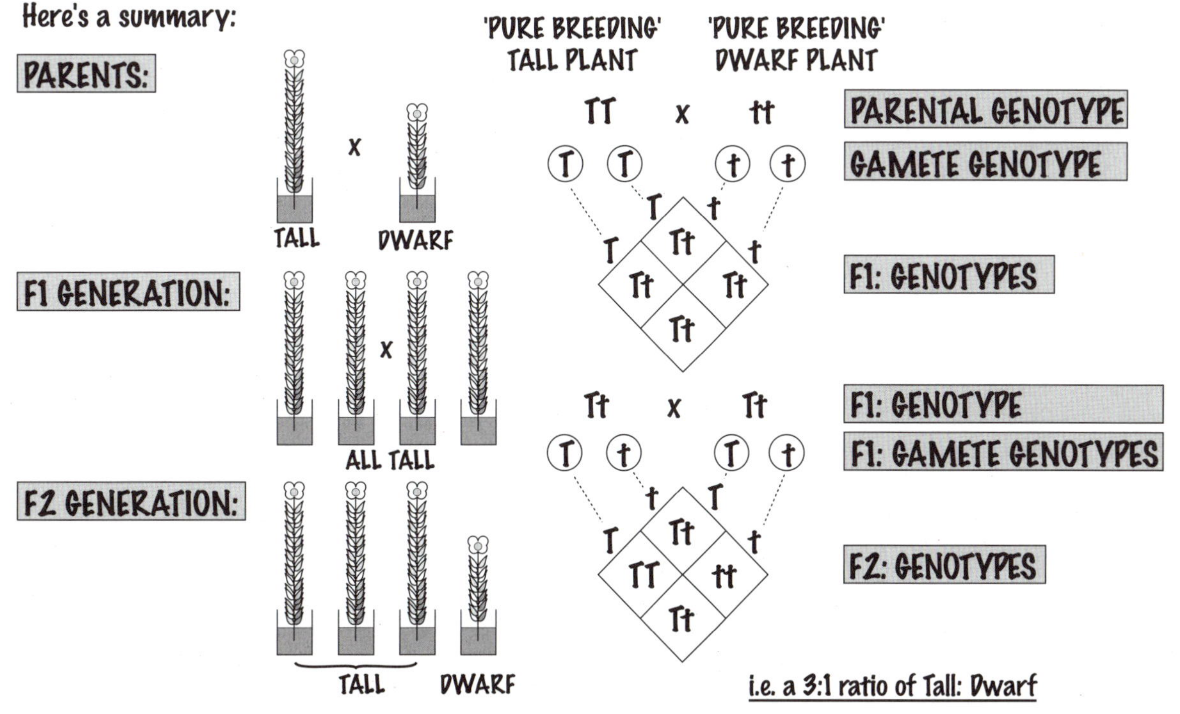

- He called "dwarfness" a recessive characteristic and interpreted the genetic patterns responsible ...
- ... However, modern microscopes and techniques hadn't been invented, and no one knew about genes and chromosomes, so his work was not recognised until after his death.

KEY POINTS:

- Genetic Engineering involves moving sections of D.N.A. from one species to another to manufacture useful biological products.

POPULATION SIZE

The SIZE of any POPULATION of PLANTS or ANIMALS within a COMMUNITY will CHANGE WITH TIME. This is due to many factors:

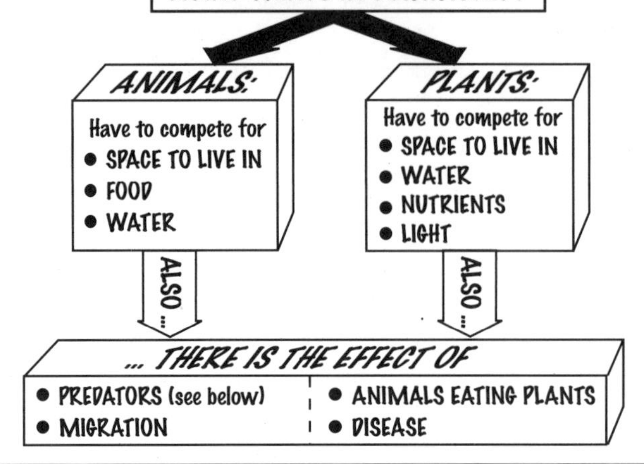

DISTRIBUTION AND ABUNDANCE OF POPULATIONS

- The DISTRIBUTION and RELATIVE ABUNDANCE of organisms in a habitat depends on ...
  ... A The ADAPTATION of the organism to its environment, ...
  ... B How much COMPETITION there is from other species ...
  ... C How much PREDATION is occurring i.e. the number and success of the predators.
- ... When two or more populations compete in a particular area or habitat, the POPULATION WHICH IS BETTER ADAPTED TO THE ENVIRONMENT IS MORE SUCCESSFUL, and usually exists in larger numbers - often resulting in the complete exclusion of the other competing populations.

ADAPTATIONS FOR SURVIVAL

- Plants and animals live, grow and reproduce in places where ...
- ... and at times when, CONDITIONS ARE SUITABLE.
- They often show obvious ADAPTATIONS ...

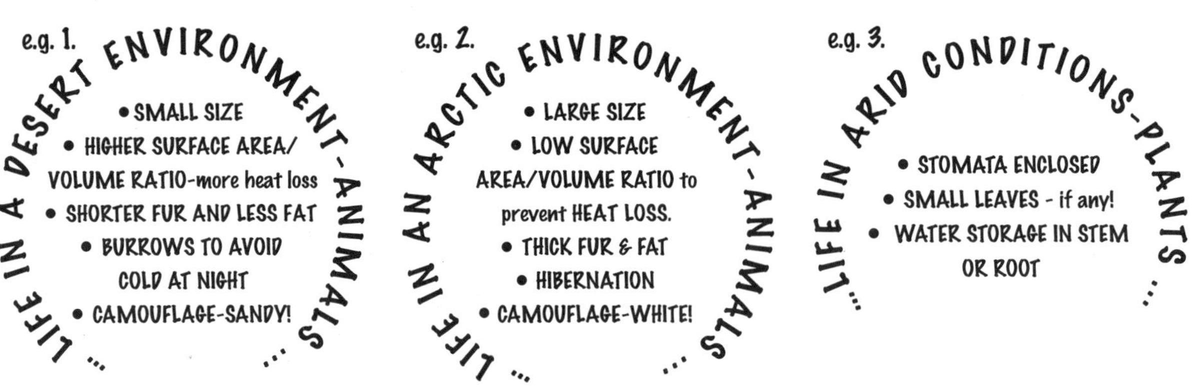

- Plants also have to be able to make the most of available ...
- ... LIGHT and SPACE.
- They show adaptations such as ...
  - CLIMBING ABILITY
  - LARGER LEAVES

KEY POINTS:

- Adaptation, Competition and Predation are three factors that affect the distribution and abundance of an organism in a habitat.

VARIATION AND EVOLUTION I - Variation

VARIATION

Differences between individuals of the **SAME SPECIES** is described as **VARIATION**.
These differences may be due to one of, or a combination of the following factors ...

- **ENVIRONMENTAL FACTORS** i.e. differences in the conditions in which they have developed.
- ... **INHERITANCE** i.e. differences in the genes they possess.
- ... **MUTATIONS** i.e. changes in their gene or chromosome structure (see MoS.8).

CONTINUOUS VARIATION - due to both Genetic and Environmental causes

If you were to measure the heights of a lot of people of the same age (say your year group), you would find ...

A • ... that **HEIGHT VARIED GRADUALLY ACROSS A RANGE** ...

B • ... that **MOST INDIVIDUALS ARE CLOSE TO THE CENTRE OF THE RANGE** (the NORM!) ...

C • ... that the **NUMBER OF INDIVIDUALS WHO DIFFER FROM THE NORM DECREASES AS THE SIZE OF THE DIFFERENCE FROM THE NORM INCREASES.**

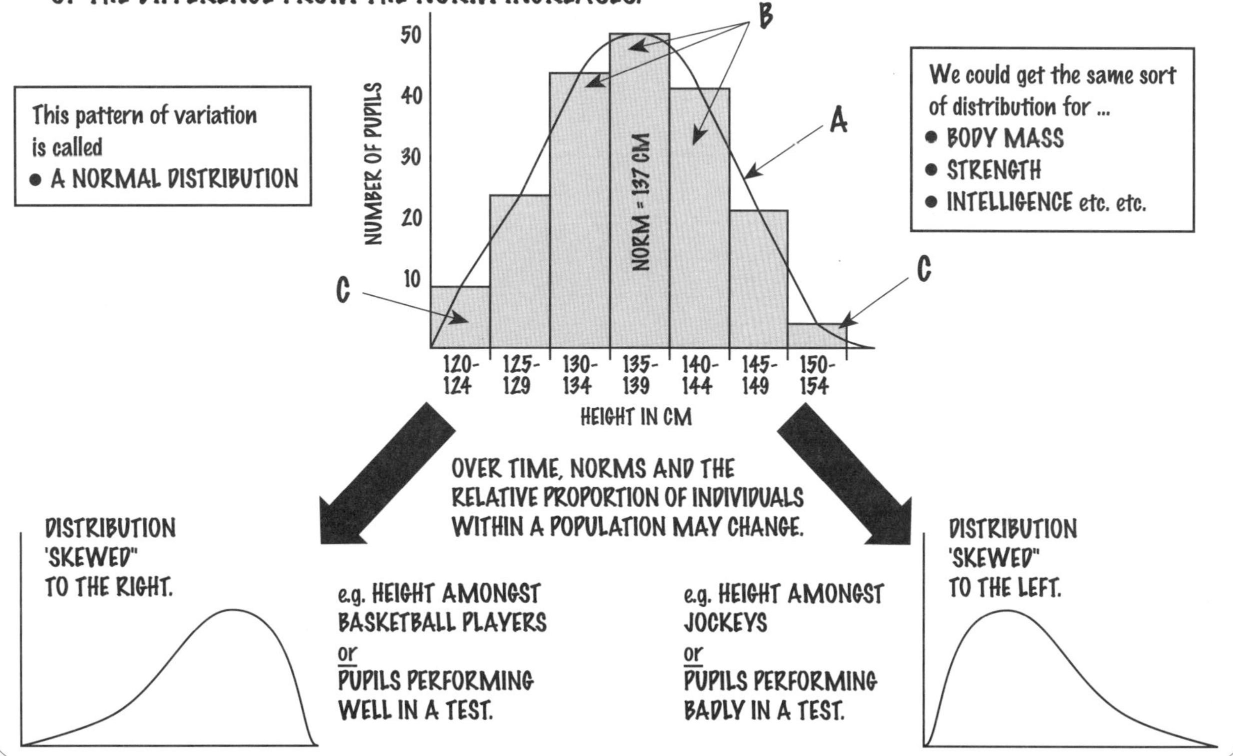

DISCONTINUOUS VARIATION - due to Genetic causes

Examples of **DISCONTINUOUS VARIATION** are ...

- **TONGUE ROLLING/NON-TONGUE ROLLING.**
- **FREE EAR LOBES/ATTACHED EAR LOBES.**
- **A/B/AB or O BLOOD GROUPS.** These are the "classic" examples, but there are many more.

The main thing to remember is that these are ...

- ... **CONTROLLED ENTIRELY BY THE GENES** and ...
- ... there are **NO HALF-MEASURES** (i.e. a person either can or can't roll his tongue)

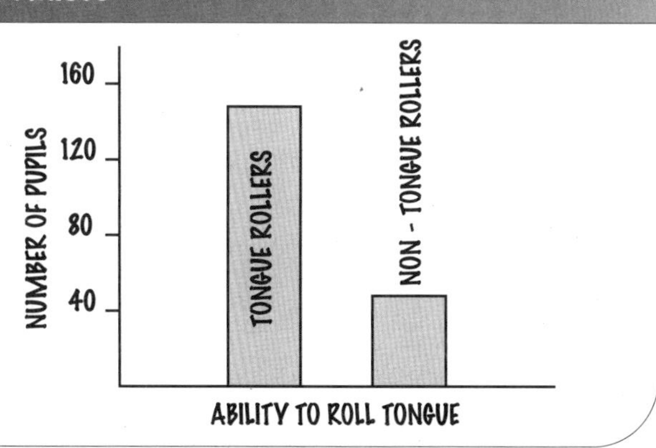

KEY POINTS:

- Continuous Variation results from both Genetic and Environmental factors.
- Discontinuous Variation results from Genetic factors only.

MUTATIONS

A MUTATION IS ...

- A major change occuring in one or more CHROMOSOMES or in the NUMBER of CHROMOSOMES ... (Large scale changes in chromosomes are usually disastrous!)
- A chemical change occuring in an INDIVIDUAL GENE, resulting in reorganisation of the D.N.A. (see MoS. 2).

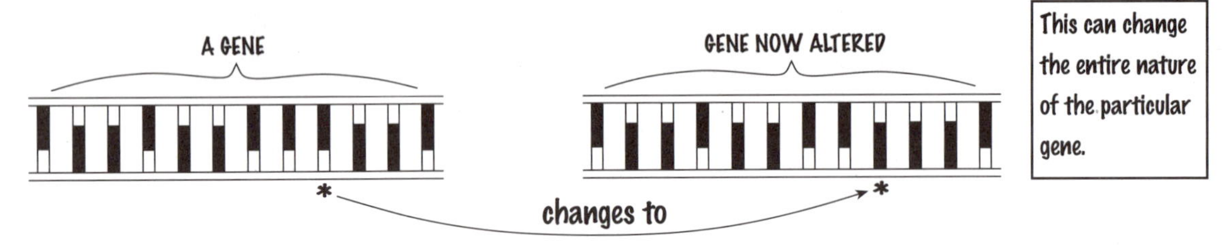

This can change the entire nature of the particular gene.

CAUSES	EFFECTS
• Mutations occur naturally but ...	• Most mutations are HARMFUL and in ...
• .... there is an increased risk of mutation if ...	• ... REPRODUCTIVE CELLS can cause DEATH or ABNORMALITY.
• ... individuals are exposed to MUTAGENS ...	• In BODY CELLS they may cause CANCER.
• ... e.g. IONISING RADIATION (inc U-V LIGHT, X-RAYS) ...	• Some mutations are NEUTRAL, and in RARE CASES ...
• ... GAMMA RAYS and CERTAIN CHEMICALS.	• ... may confer a SURVIVAL ADVANTAGE, on an organism ...
• THE GREATER THE DOSE, THE GREATER THE RISK.	• ... and its OFFSPRING WHO INHERIT THE GENE (MoS.9).

SELECTIVE BREEDING

- Selective breeding or ARTIFICIAL SELECTION means taking an organism with a DESIRED CHARACTERISTIC ...
- ... and CROSS-BREEDING it with another of the SAME SPECIES which has the DESIRED CHARACTERISTIC ...
- ... in the hope that some of the offspring will have an EXAGGERATED VERSION OF THIS CHARACTERISTIC.

This has been done with COWS, SHEEP, PIGS, HENS, WHEAT, CABBAGES, SPROUTS, CAULIFLOWER etc.
In the case of plants, once the desired result has been achieved ...
... many individuals can be produced by SELF-POLLINATION.

EXAMPLE OF GREEN VEGETABLES BRED FROM A COMMON ANCESTOR

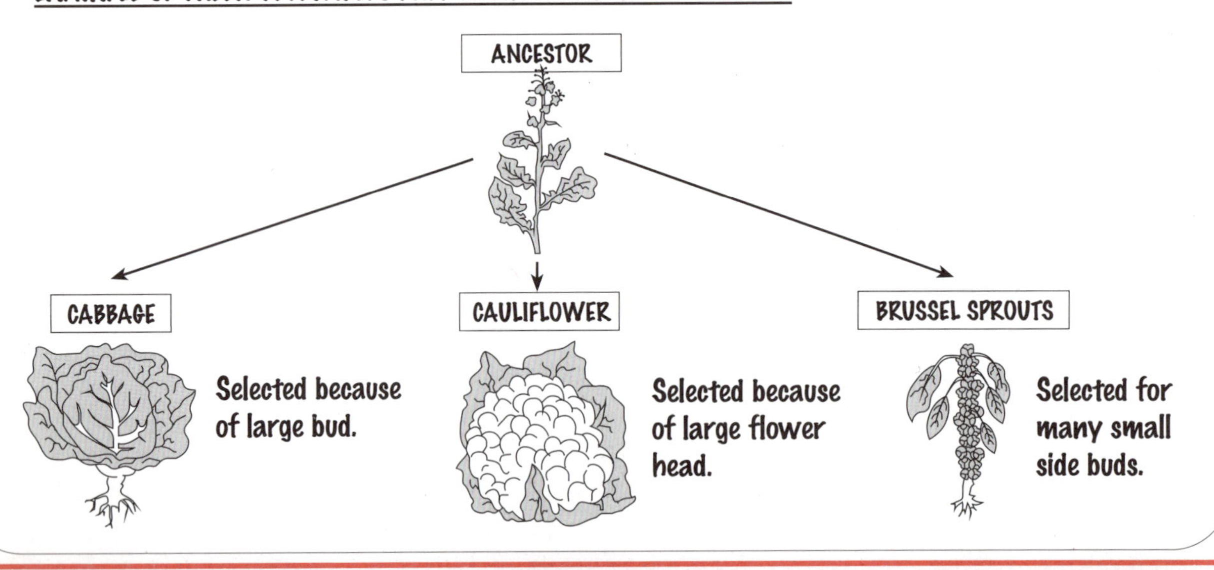

KEY POINTS:

- A change occurring in one or more chromosomes is a Mutation.
- Selective breeding is the cross-breeding of two organisms of the same species with a view to producing a desired characteristic.

SURVIVAL

- The reassortment of genetic material in **SEXUALLY REPRODUCING ORGANISMS** promotes **VARIATION**.
- Mutations, in organisms which reproduce **SEXUALLY** or **ASEXUALLY** also promotes **VARIATION**.
  Sometimes variation confers a **SURVIVAL ADVANTAGE** on particular individuals which ...
  ... particularly in the case of a **CHANGING ENVIRONMENT** may make them **BETTER ADAPTED**.

NATURAL SELECTION

Evolution is the change in a population over a large number of generations ...
... that may result in the formation of a new species; which are better adapted to their environment.

There are 4 key points to remember:-

1. Individuals within a population show **VARIATION** (i.e. differences due to their genes).

2. There is **COMPETITION** between individuals for food and mates etc., and also predation and disease. This keeps population sizes constant in spite of the production of many offspring, i.e. there is a "struggle for survival".

3. Some individuals (because of **VARIATION**) may have a **SURVIVAL ADVANTAGE** which makes them **BETTER ADAPTED** to their environment and consequently more likely to survive, breed successfully and produce offspring.

4. These 'survivors' will therefore **PASS ON THEIR GENES** to these offspring resulting in an improved organism being evolved through **NATURAL SELECTION**.

The Classic Example - The Peppered moth

VARIATION — and →	COMPETITION — ensure →	BETTER ADAPTED →	PASS ON THEIR GENES
Moths of different shades rest on "sooty" silver birch trees.	The lighter ones stand out and are eaten by birds.	Darker ones survive and pass on their genes.	Population in urban areas become uniformly darker.

The dark variation has a **SURVIVAL ADVANTAGE** in times of **CHANGING ENVIRONMENT**.

Some strains of **BACTERIA** are **RESISTANT TO PENICILLIN** for exactly the same reasons, as those outlined above. The antibiotic (penicillin) acts as the agent for selection resulting in the survival of **RESISTANT BACTERIA**.

CONFLICTING THEORIES OF EVOLUTION

1. **CHARLES DARWIN (1809-1882)** proposed the **THEORY OF NATURAL SELECTION** and suggested that species evolve due to inheritance of small changes which have passed the survival test of natural selection.

2. **LAMARCK (1744-1829)** suggested that species evolve due to inheritance of changes acquired by the use or disuse of various body parts i.e. Giraffes evolved longer necks by constantly straining to reach up to higher leaves, and humans 'lost their tail' by failing to use it enough!!

- We now know that **DARWIN** was referring to changes in the **GENOTYPE** and **LAMARCK** the **PHENOTYPE**, and of course organisms **INHERIT GENOTYPES** not **PHENOTYPES**.
- These different interpretations and the general hostility of the church clouded the issues for many years.

KEY POINTS:

- Evolution is the change in a population over a large number of generations.
- Charles Darwin and Lamarck provided conflicting theories of evolution. Darwin was right.

SPECIES

- Organisms which belong to the SAME SPECIES can freely INTERBREED ...
- ... to produce FERTILE OFFSPRING.
  This last bit is important for while a horse and a donkey can interbreed, the offspring are STERILE.

FOSSIL RECORDS

- Fossils are the "remains" of plants or animals from many years ago which are found in rocks.
- They are preserved usually by being covered in sediment which causes them "to turn to stone" ...
- ... or because there is an absence of decay conditions where they have fallen to rest.
- In sedimentary rock, the deeper the layer the older the fossils (usually!).

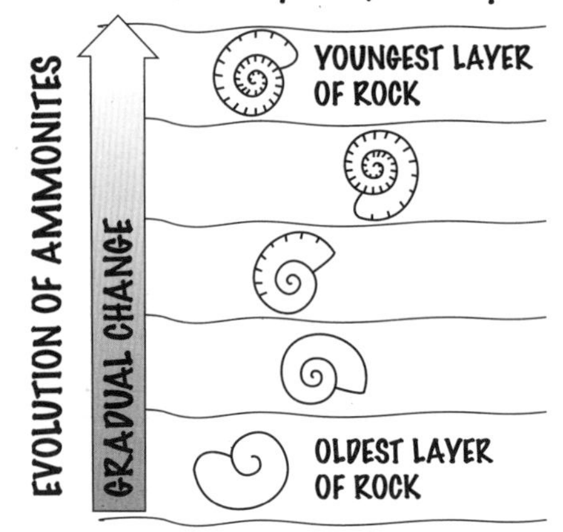

- If we look at exposed rock strata, ...
- ... it is possible to follow the GRADUAL CHANGES which have taken place in an organism over time.
- Even though the fossil record is incomplete, these gradual changes confirm that ...
- ... SPECIES HAVE CHANGED OVER LONG PERIODS OF TIME.
- ... providing strong evidence for evolution.

EXTINCTION

- If a species is EXTINCT, it means there are no living individuals of that species left on the planet.
- Fossil records and documented natural history shows that many species have become EXTINCT ...
- ... often because they have failed to solve the biological problems set by a change in the environment.
- i.e. they have been unable to adapt quickly enough.

EVIDENCE FOR EXTINCTION PROVIDED BY THE FOSSIL RECORD includes the DINOSAURS.
- Dinosaur fossils are now common place and reveal the diversity of this group.
- So far no preserved specimens have come to light and are unlikely to do so.
- There are many theories to account for their extinction but whichever you accept ...
- ... they are an EVOLUTIONARY "DEAD-END". (i.e. They were "selected against.")

EVIDENCE FOR EXTINCTION PROVIDED BY NATURAL HISTORY SPECIMENS includes MAMMOTHS.
- Mammoths (similar to elephants) have been discovered ...
- ... preserved in ICE in SIBERIA, and TAR PITS in other parts of the world ...
- ... revealing them to be examples of an extinct species.
- Examples are on display in various museums around the world ...
- ... of yet another EVOLUTIONARY "DEAD END". (i.e. They were "selected against.")

KEY POINTS:

- Fossils are the preserved remains of plants and animals.
- Evidence for extinction is provided by the Fossil Record and Natural History specimens.

POLLUTION FROM FOSSIL FUELS

Most air pollution is caused by the BURNING OF FOSSIL FUELS. The main sources of gases are:
• INDUSTRY • POWER STATIONS • MOTOR VEHICLE EXHAUSTS

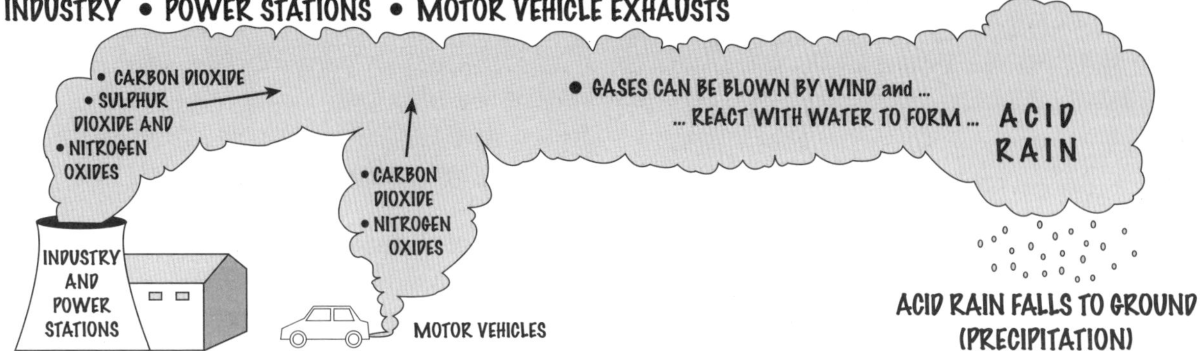

- ACID RAIN damages TREES and raises the ACIDITY of LAKES and RIVERS killing FISH and other wildlife.
- Emissions of CARBON DIOXIDE contribute to GLOBAL WARMING (see below).

DEFORESTATION

Demand for hardwoods and need for land has resulted in worldwide destruction of the rainforests.

- This has INCREASED the CARBON DIOXIDE CONTENT of the atmosphere (less photosynthesis.)
- REDUCED TRANSPIRATION has led to REDUCTION in RAINFALL.
- Newly exposed land is now vulnerable to SOIL EROSION as water flows over it removing TOP SOIL.

GREENHOUSE EFFECT AND GLOBAL WARMING

- The Greenhouse effect relies on CARBON DIOXIDE and METHANE absorbing heat in the atmosphere, preventing it RADIATING OUT INTO SPACE.
- This, together with the increase in the amount of waste heat generated on the planet has resulted in ... ... GLOBAL WARMING, which is now an international problem.

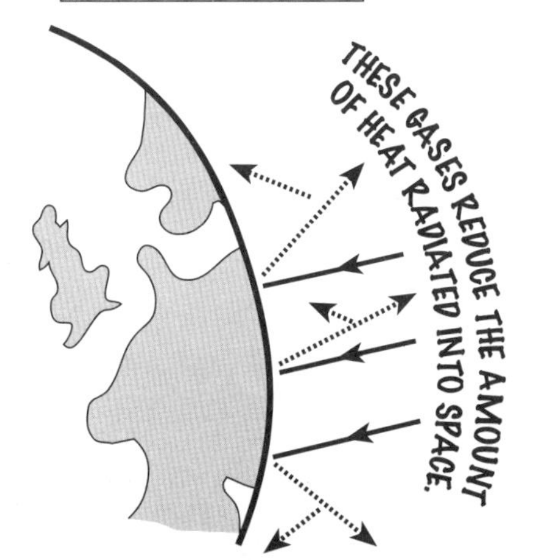

BURNING FOSSIL FUELS - carbon dioxide
+
DEFORESTATION - carbon dioxide
+
CATTLE, RICE FIELDS, MICROBES - methane

cause
GLOBAL WARMING

CFC's AND THE OZONE LAYER

- The OZONE LAYER absorbs HIGH ENERGY, ULTRA-VIOLET RADIATION from the SUN.
- This HIGH ENERGY RADIATION is dangerous and prolonged exposure can cause SKIN CANCER.
- C.F.C.'s (chlorofluorocarbons) are used as PROPELLANTS in AEROSOLS, and in REFIGERANTS ...
- ... but unfortunately they BREAKDOWN THE OZONE LAYER when released into the atmosphere ...
- ... resulting in HIGH ENERGY, U-V LIGHT reaching the surface of the planet!!
- There is now a 'hole' in the ozone layer, ... and an increased incidence of skin cancer.

KEY POINTS:

- Burning Fossil Fuels results in air pollution creating Acid Rain.
- Global Warming is due to the Greenhouse Effect.
- Breakdown of the Ozone layer is due to C.F.C.'s.

Food production involves the management of ecosystems to improve the efficiency of energy transfer from the sun to human food. Such management imposes a duty of care for the environment, in order to avoid the destruction of ecosystems ..

DAMAGING ECOSYSTEMS

Methods of food production which have damaged ecosystems include:-

USE OF PESTICIDES
- Widespread use of pesticides leads to ...
- ... WHOLESALE, INDISCRIMINATE, DESTRUCTION OF INSECTS, causing ...
- ... BREAKS IN THE FOOD CHAIN ...
- ... ACCUMULATION OF PESTICIDES IN ORGANISMS HIGHER UP THE FOOD CHAIN ...
- ... and REDUCTION IN THE NUMBER OF POLLINATING INSECTS.

USE OF INORGANIC FERTILISERS
- Widespread use of inorganic fertiliser leads to ...
- ... large amounts being washed off the land into water courses.
- This LEACHING EFFECT results in EUTROPHICATION of STREAMS and LAKES, ... (see below)
- ... resulting in REDUCED OXYGEN CONTENT of the water, ...
- ... and the DEATH OF LARGE NUMBERS OF AQUATIC ORGANISMS.

USE OF LARGER FIELDS
- Results in the REMOVAL OF MILES OF NATURAL HABITAT e.g. HEDGES.
- The loss of the hedges REMOVES THE NATURAL WINDBREAKS, ...
- ... causing TOPSOIL TO BE GRADUALLY BLOWN OFF THE LAND.

MANAGING ECOSYSTEMS

It is possible to have efficient food production and maintain a balanced ecosystem. In addition to the 'solutions' below, hard decisions have to be made taking into account ...
... • HISTORICAL, ... • POLITICAL, ... • ECONOMIC, ... and SOCIAL ISSUES.

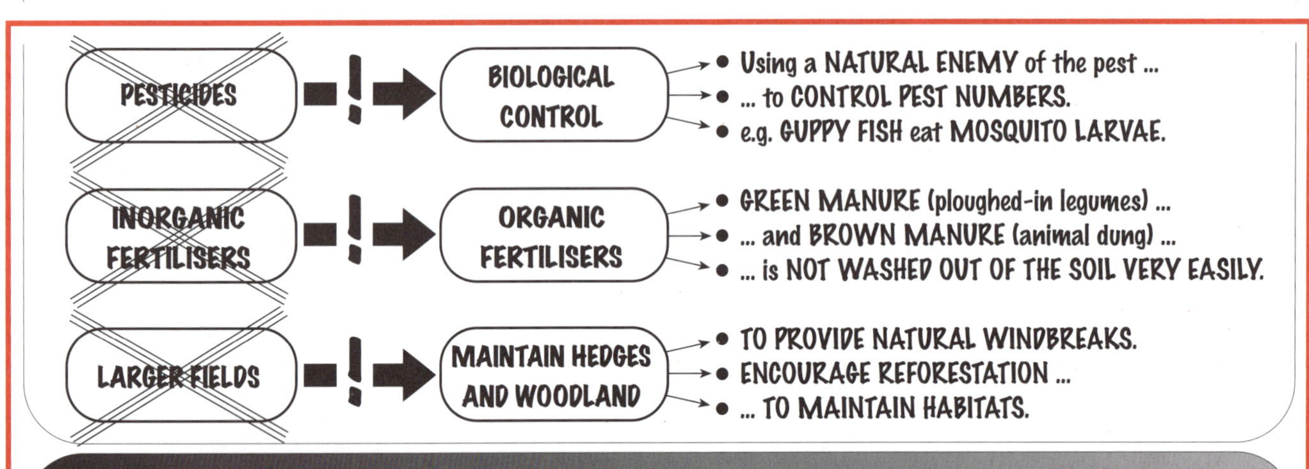

EUTROPHICATION - The six stages

1. INORGANIC FERTILISERS ... used by farmers may be washed into lakes and rivers.
2. GROWTH ... of water plants caused by this fertiliser, happens rapidly.
3. DEATH ... of some of these plants due to lack of light from overcrowding.
4. MICROBES ... which feed on dead organisms now increase massively in number.
5. OXYGEN ... is used up quickly by this huge number of microbes (as they respire)
6. SUFFOCATION ... of fishes and other aquatic animals due to lack of oxygen in the water.

KEY POINTS:

- Ecosystems are damaged through the use of Pesticides, Inorganic Fertilisers and Larger Fields.

USE AND MIS-USE OF DRUGS

Drugs are substances which alter the natural chemistry of the body ...
... causing it to work differently.

HELPFUL DRUGS

1. PAINKILLERS (ANALGESICS) - Have an effect on either the PERIPHERAL ...
   ... or CENTRAL NERVOUS SYSTEM by affecting the transmission or reception of nerve impulses.
2. ANTIBIOTICS - are substances produced by bacteria or fungi, and extracted by man. e.g. penicillin.
   They are used to kill micro organisms which invade the human body.

DANGEROUS DRUGS

Any drug which is mis-used may be dangerous, but especially ...

1. ALCOHOL which causes ...
   - impairment of judgement ...
   - ... slower reactions and ...
   - ... damage to brain and liver cells.
2. SOLVENTS which cause ...
   - hallucinations ("seeing things") and ...
   - ... damage to brain, liver and kidneys.
3. TOBACCO which may cause ...
   - temporary damage to the ciliated epithelium (see below) ...
   - ... causing tar to become deposited which ...
   - ... may cause cancer.
   - Addiction to nicotine is also very difficult to break.

OTHER EFFECTS OF TOBACCO

A mucous membrane is the name given to a MOIST EPITHELIUM (covering layer).

- This type of epithelium is present in the 'breathing passages", and has CILIA to help 'beat' the mucus upwards to the mouth.
- It helps to keep the air MOIST and WARM.

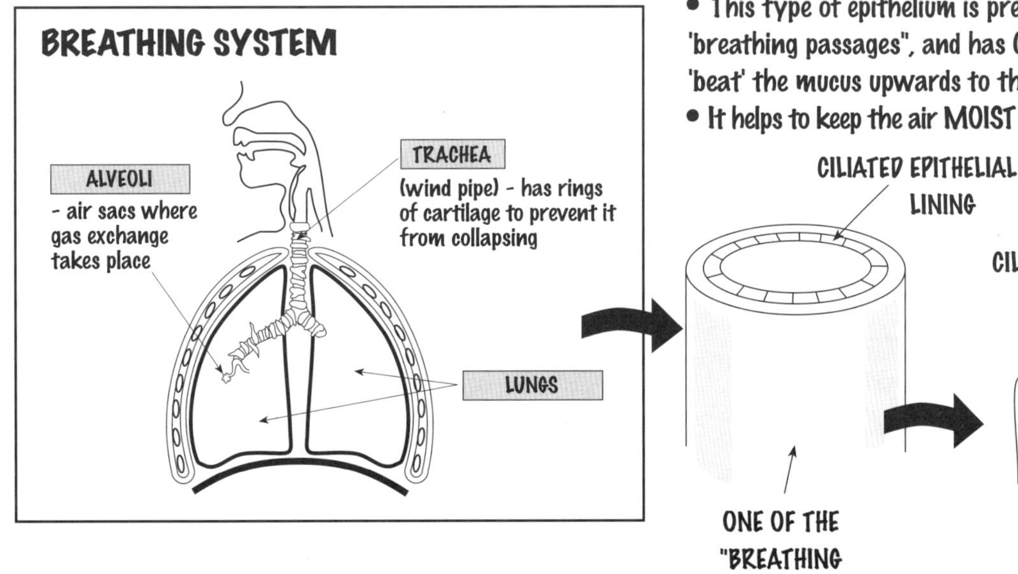

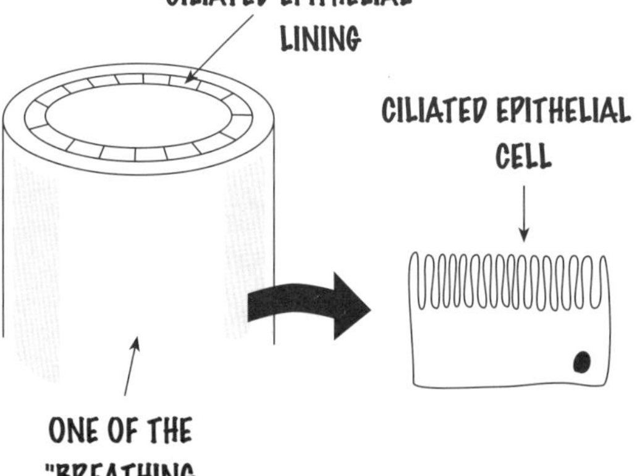

- Tobacco smoke causes the CILIA to stop propelling MUCUS up the breathing tubes to the mouth.
- Tobacco may also cause LOW BIRTH WEIGHT in foetus because of lack of oxygen, ...
- ... caused by CARBON MONOXIDE REDUCING THE OXYGEN CARRYING CAPACITY OF THE BLOOD.

KEY POINTS:

- Drugs alter the natural chemistry of the body.
- Painkillers and Antibiotics are helpful while Alcohol, Solvents and Tobacco are dangerous.

1. Describe what happens in sexual reproduction.
2. Describe what happens in asexual reproduction.
3. What happens to chromosome number during meiosis?
4. What happens to chromosome number during mitosis?
5. Describe what is meant by a gene.
6. What is the difference between a gene and a chromosome.
7. What is an allele?
8. What is meant by a) homozygous and b) heterozygous?
9. What is the difference between genotype and phenotype?
10. Draw a genetic diagram to show how gender is determined.
11. What is monohybrid inheritance?
12. Describe the difference between parental, F1 and F2 generations.
13. Describe the systems of Cystic Fibrosis.
14. Draw a genetic diagram of a cross between two carriers for this condition.
15. Describe the systems of Sickle Cell Anaemia.
16. Draw a genetic diagram of a cross between two carriers for this condition.
17. Give an example of genetic engineering.
18. Draw a genetic diagram showing Mendel's early work on pea plants.
19. Which factors affect the size and distribution of populations?
20. Give three examples of animals and their adaptations to their environment.
21. What is variation?
22. What may cause it?
23. Draw a graph of distribution of height across a year group.
24. Describe what is meant by the term mutation.
25. What may cause mutations?
26. What is selective breeding?
27. What are the key factors involved in natural selection?
28. Give an example of natural selection.
29. What are the main differences between the theories of Darwin and Lamarck?
30. Define species.
31. How does the fossil record provide evidence for evolution?
32. What is meant by the word extinct?
33. Give two examples of extinction.
34. Explain why these happened.
35. What are the effects of deforestation?
36. What is the greenhouse effect?
37. What does the ozone layer do?
38. What is causing it to break down?
39. Give three examples of methods of food production which have damaged ecosystems.
40. Describe possible remedies for each of these.
41. Describe the six stages of eutrophication.
42. Give an example of a helpful drug.
43. Give an example of a dangerous drug.
44. Describe the effects of this dangerous drug.
45. What effect does tobacco have on the breathing passages?
46. Describe any other effects of tobacco.

STRUCTURE AND CHANGES

ATOMIC STRUCTURE SC 1

Originally ATOMS WERE considered to be the SMALLEST PARTICLES OF MATTER, however they are NOW considered to CONSIST OF ...

- ... a small CENTRAL NUCLEUS ...
- ... made up of PROTONS and NEUTRONS (one exception!) ...
- ... surrounded by ELECTRONS arranged in SHELLS.

A SIMPLE EXAMPLE - Helium

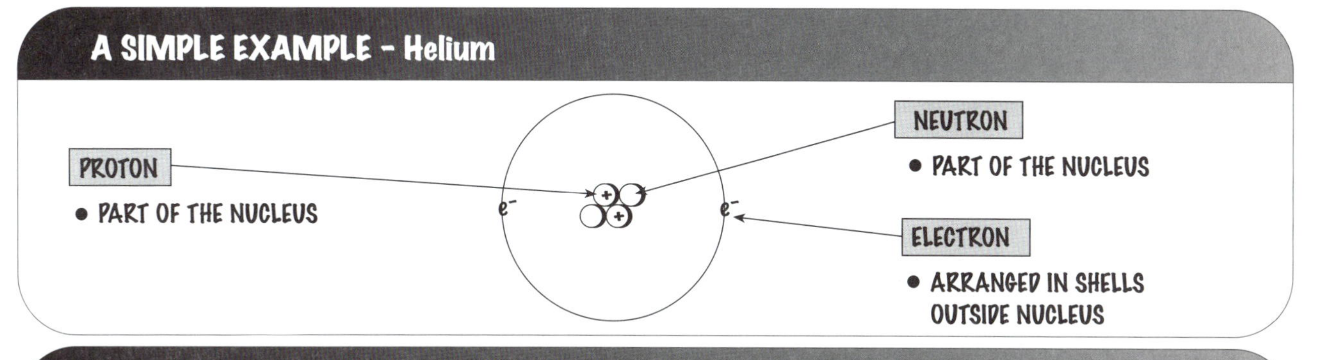

ATOMIC NUMBER

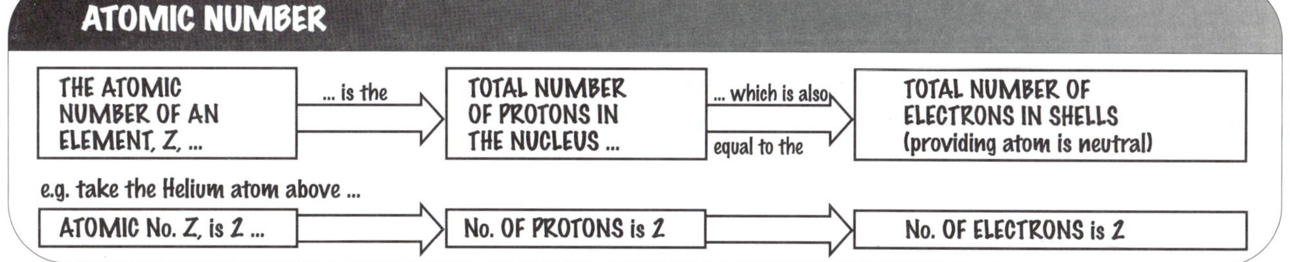

ARRANGEMENT OF ELECTRONS

The electrons in an atom occupy the INNERMOST AVAILABLE SHELLS (Lowest available energy levels)

- The first shell can contain a MAXIMUM OF TWO ELECTRONS.
- All subsequent shells contain a MAXIMUM OF EIGHT ELECTRONS.
- The NUMBER OF ELECTRONS IN THE OUTERMOST SHELL IS EQUAL TO THE PERIODIC TABLE GROUP No.

ELEMENT	Hydrogen, H	Helium, He	Lithium, Li	Beryllium, Be
ATOMIC NO, Z	1	2	3	4
No. OF ELECTRONS	1	2	3	4
ELECTRON CONFIGURATION	1.	2.	2.1	2.2
PERIODIC TABLE GROUP No.	1	2	1	2
'DOT-AND-CROSS' DIAGRAM (electrons are represented by either 'dots' or 'crosses')				

ELEMENT	Boron, B	Oxygen, O	Silicon, Si	Argon, Ar
ATOMIC NO, Z	5	8	14	18
No. OF ELECTRONS	5	8	14	18
ELECTRON CONFIGURATION	2.3	2.6	2.8.4	2.8.8
PERIODIC TABLE GROUP No.	3	6	4	0 (or 8)
'DOT-AND-CROSS' DIAGRAM				

KEY POINTS:

- An atom consists of a small central nucleus made up of protons and neutrons surrounded by electrons arranged in shells.
- The atomic number of an element is the total number of protons in the nucleus. • Electrons are arranged in shells where the innermost shell contains a maximum of two electrons. All subsequent shells contain a maximum of eight electrons.

THE PERIODIC TABLE

The 90 naturally occurring elements are grouped into "FAMILIES" WITH SIMILAR PROPERTIES.
These "FAMILIES" of elements are then arranged into the PERIODIC TABLE.

← GROUPS →

PERIODS	1	2		3	4	5	6	7	0 (or 8)
1			H 1	NON-METALS					He 2
2	Li 3	Be 4		B 5	C 6	N 7	O 8	F 9	Ne 10
3	Na 11	Mg 12		Al 13				Cl 17	Ar 18
4	K 19	Ca 20	Fe 26, Ni 28, Cu 29, Zn 30					Br 35	Kr 36
5			TRANSITION METALS: Ag 47					I 53	Xe 54
6			Pt 78, Au 79, Hg 80		Pb 82				
7									

(For the full table, see the inside back cover)

KEY POINTS:

- The FIRST periodic table by John Newlands, 1864, arranged the known elements (only 63 of them!) in order of increasing RELATIVE ATOMIC MASS.
- The Elements are now arranged in order of INCREASING ATOMIC NUMBER (see diagram).
- Elements in the SAME GROUP have SAME NUMBER OF ELECTRONS IN THEIR OUTERMOST SHELL ...
  ... (this number also coincides with the GROUP NUMBER) ...
  ... and have SIMILAR PROPERTIES.
- From left to right ACROSS EACH PERIOD, A SHELL IS GRADUALLY FILLED WITH ELECTRONS ...
  ... in the next period, the next shell is filled.

GROUP 0 - THE NOBLE GASES (sometimes called group 8)

He Helium
Ne Neon
Ar Argon

There are SIX GASES in this group, but the top three in the group are the ones we need to deal with.

- **They exist as INDIVIDUAL ATOMS (Monatomic)**
  e.g. He, Kr etc. rather than diatomic gases like other gaseous elements (Cl_2, H_2 etc).
- **They are CHEMICALLY UNREACTIVE**
  All these elements have a "fully occupied" outer shell (energy level) e.g. Neon is 2,8 and Argon is 2,8,8 ...
  i.e. their outer shells are full.
  Because of this they tend not to 'want' to gain, lose, or share electrons and therefore they are UNREACTIVE.
- **USES**
  Helium is used to fill airships. Neon in electrical discharge tubes to give colour in advertising signs. Argon as the inert gas in light bulbs

Kr Krypton
Xe Xenon
Rn Radon

KEY POINTS:

- The Periodic Table is a list of elements arranged in order of increasing atomic number.
- Elements in the same group have similar properties since they have the same number of electrons in their outermost shell.
- Group 0 elements, known as the Noble Gases, exist as individual atoms and are chemically unreactive.

GROUP I - THE ALKALI METALS

1
Li Lithium (Z = 3)
Na Sodium (Z = 11)
K Potassium (Z = 19)

more reactive

There are SIX METALS in the group, but the top three are the ones we need to deal with:

- They have a LOW MELTING POINT and LOW DENSITY
- Their REACTIVITY INCREASES with INCREASING ATOMIC NUMBER, Z
- They REACT with COLD WATER FORMING METAL HYDROXIDES and RELEASING HYDROGEN

e.g. $2K_{(s)} + 2H_2O_{(l)} \longrightarrow 2KOH_{(aq)} + H_{2(g)}$

Potassium + water ⟶ Potassium hydroxide + Hydrogen

As we move down the group, the metals react more VIGOROUSLY with water ... ... they float, may melt, ... and the hydrogen gas may ignite!

- The ALKALI METAL HYDROXIDES so formed dissolve in WATER to give ALKALINE SOLUTIONS.

i.e. pH of solution is GREATER THAN 7.

- ALKALI METAL HALIDES (e.g. NaCl) ALSO dissolve in WATER

REACTION WITH OXYGEN AND HALOGENS TO FORM IONIC SOLIDS

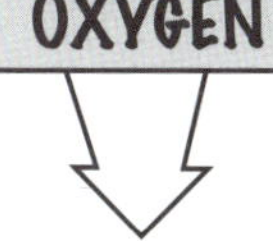

OXYGEN

- Metal atom loses ONE electron to form 1^+ ION (Li^+)
- Oxygen atom gains TWO electrons to form 2^- ION (O^{2-})

e.g. $4Li_{(s)} + O_{2(g)} \longrightarrow 2Li_2O_{(s)}$

Lithium + Oxygen ⟶ Lithium oxide

HALOGEN

- Metal atom again loses ONE electron to form 1^+ ION (Na^+)
- Halogen atom gains ONE electron only to form 1^- ION (Cl^-)

e.g. $2Na_{(s)} + Cl_{2(g)} \longrightarrow 2NaCl_{(s)}$

Sodium + Chlorine ⟶ Sodium chloride

ALKALI METAL COMPOUNDS

Both Alkali metal hydroxides and halides are IONIC compounds which separate out into their IONS when dissolved in water.

HYDROXIDE

- e.g. Sodium hydroxide, Na OH ...
- ... forms $Na^+_{(aq)}$ and $OH^-_{(aq)}$ IONS.
- ALL ALKALINE SOLUTIONS CONTAIN $OH^-_{(aq)}$ IONS
- This is why they are called the ALKALI METALS.

HALIDE

- e.g. Sodium chloride, NaCl
- ... forms $Na^+_{(aq)}$ and $Cl^-_{(aq)}$ IONS.

ELECTROLYSIS OF SODIUM CHLORIDE SOLUTION

As we have seen Sodium chloride (common salt) is a compound of an ALKALI METAL and a HALOGEN. Electrolysis of sodium chloride solution produces CHLORINE, SODIUM HYDROXIDE SOLUTION and HYDROGEN, all of which have a great number of uses.

$$2NaCl_{(aq)} + 2H_2O_{(l)} \longrightarrow Cl_{2(g)} + 2NaOH_{(aq)} + H_{2(g)}$$

Sodium chloride + water ⟶ Chlorine + Sodium hydroxide + Hydrogen

KEY POINTS:

- Group 1 elements are known as the Alkali Metals.
- Their reactivity increases as we go down the group.
- They have a low melting point and low density.

GROUP 7 - THE HALOGENS

7
F Fluorine (Z = 9)
Cl Chlorine (Z = 17)
Br Bromine (Z = 35)
I Iodine (Z = 53)

less reactive

There are FIVE NON-METALS in this group, but the only ones we need to be concerned about are chlorine, bromine and iodine

- They all have COLOURED VAPOURS which in the case of chlorine and bromine are extremely pungent.
- Their MELTING POINT and BOILING POINT INCREASE with INCREASING ATOMIC No. (Z)
- Their REACTIVITY DECREASES with INCREASING ATOMIC No. Z
- HYDROGEN HALIDES (e.g. HCl) form ACIDIC SOLUTIONS when added to WATER.

i.e. pH of solution is LESS THAN 7.

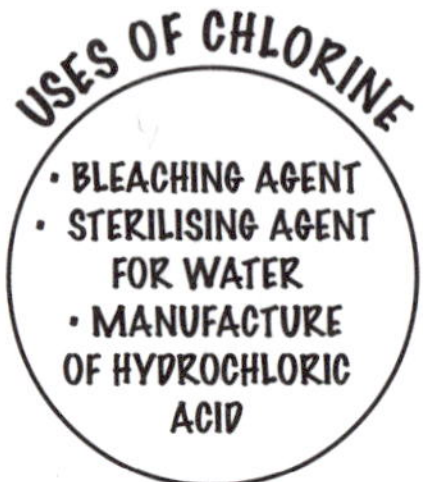

REACTION WITH ALKALI METALS AND HYDROGEN

ALKALI METAL

- Forms solid IONIC COMPOUNDS (see SC.3) with the ...
- ... HALIDE ION (chloride, bromide or iodide ion) ...
- ... having a 1^- charge.
- Metal has 1^+ charge.

e.g. $2K_{(s)} + Br_{2\,(g)} \longrightarrow 2KBr_{(s)}$

Potassium + Bromine ⟶ Potassium bromide

HYDROGEN

- Forms COVALENT COMPOUNDS where ...
- ... HALOGEN and HYDROGEN 'share' electrons
- e.g. $Cl_{2\,(g)} + H_{2\,(g)} \longrightarrow 2\,HCl_{(g)}$

Chlorine + Hydrogen ⟶ Hydrogen chloride.

- When added to water HYDROGEN HALIDES ...
- ... form ACIDIC SOLUTIONS as there are H^+ IONS present

e.g. HCl forms H^+ (aq) (and Cl^-(aq) ions).

DISPLACEMENT REACTIONS BETWEEN TWO HALOGENS

- A MORE REACTIVE halogen will DISPLACE a LESS REACTIVE halogen and so CHLORINE will DISPLACE BOTH BROMINE and IODINE from solution while BROMINE will DISPLACE IODINE.

e.g. $2KI_{(aq)} + Cl_{2(g)} \longrightarrow 2KCl_{(aq)} + I_{2\,(aq)}$

Potassium iodide + chlorine ⟶ Potassium chloride + iodine

- You will have noticed from all the reactions above that the HALOGENS exist as MOLECULES i.e. PAIRS OF ATOMS (diatomic)

ALKALI METALS AND HALOGENS - HOW ATOMIC NUMBER AFFECTS THEIR PROPERTIES

ALKALI METALS

Li
Na
K

more reactive; lower melting and boiling points

AS THE ATOMIC NUMBER INCREASES ...

- MELTING POINT AND BOILING POINT DECREASES.
- REACTIVITY INCREASES as OUTER ELECTRON SHELL increases in distance from the nucleus. Easier for electrons to be lost from outer shell as the PULLING FORCE of the NUCLEUS is DECREASED.

- MELTING POINT AND BOILING POINT INCREASES.
- REACTIVITY DECREASES as OUTER ELECTRON SHELL increases in distance from the nucleus. Harder for electrons to be added to outer shell as the PULLING FORCE of the NUCLEUS is DECREASED.

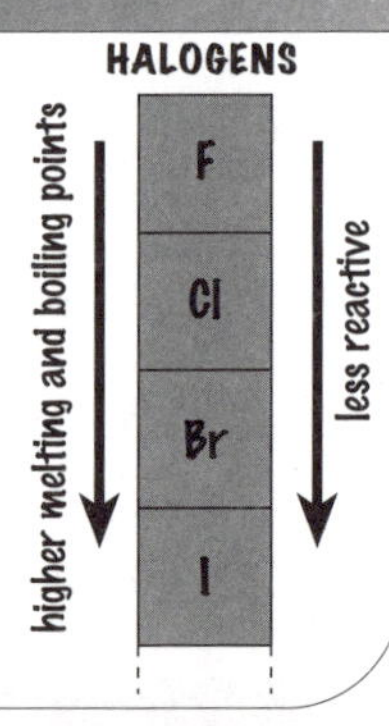

KEY POINTS:

- Group 7 elements, the Halogens, have coloured vapours.
- Their reactivity decreases as we go down the group.

REACTIVITY SERIES 1 - The reactivity series

Some metals willingly combine with other elements to form new COMPOUNDS. By observing how metals react with OXYGEN, WATER, and DILUTE ACIDS we can place them in order of their reactivity. We call this list the REACTIVITY SERIES:

	SYMBOL	ELEMENT	REACTION WITH OXYGEN	REACTION WITH WATER	REACTION WITH DILUTE HYDROCHLORIC ACID OR SULPHURIC ACID
And this is it! VERY REACTIVE	K	Potassium	Burn very easily with a bright flame	Burns violently in cold water	Violent reaction (very dangerous)
VERY REACTIVE	Na	Sodium		Vigorous reaction in cold water	
VERY REACTIVE	Ca	Calcium		Reacts slowly in cold water	
QUITE REACTIVE	Mg	Magnesium		Slowly in cold water V. fast in steam	Reasonable reaction becoming slower down the list
QUITE REACTIVE	Al	Aluminium	React slowly when heated	Layer of oxide stops reaction	
QUITE REACTIVE	*1 Zn	Zinc		Reacts quickly in steam	
NOT SO REACTIVE	Fe	Iron		Reacts reversibly with steam	
NOT SO REACTIVE	*2 Cu	Copper		No reaction with water or steam	No reaction
NOT REACTIVE AT ALL	Au	Gold	No reaction		

INCREASING REACTIVITY (upwards)

*1 POSITION OF CARBON *2 POSITION OF HYDROGEN

We also need to know what is produced when these metals react with OXYGEN, WATER and ACID:-

REACTION WITH AIR (OXYGEN) Nearly all the metals react with OXYGEN to form OXIDES.

METAL + OXYGEN ⟶ METAL OXIDE
e.g. Iron + Oxygen ⟶ Iron Oxide

When a metal is dull or TARNISHED it is because it has a layer of oxide on it.

REACTION WITH COLD WATER Some metals react chemically producing HYDROXIDES and HYDROGEN.

METAL + WATER ⟶ METAL HYDROXIDE + HYDROGEN
e.g. Sodium + Water ⟶ Sodium Hydroxide + Hydrogen

Potassium and Calcium also react like this.
We know its Hydrogen if it 'POPS' when held near a lighted splint.

REACTION WITH DILUTE ACIDS Many metals react with acids to produce a SALT and HYDROGEN

METAL + ACID ⟶ SALT + HYDROGEN
e.g. Magnesium + Hydrochloric acid ⟶ Magnesium Chloride + Hydrogen
e.g. Magnesium + Sulphuric acid ⟶ Magnesium Sulphate + Hydrogen

Different salts are made from different acids. METAL CHLORIDES from HYDROCHLORIC ACID, and METAL SULPHATES from SULPHURIC ACID.
(N.B. The only reactions with acids you need to know are for magnesium, aluminium, zinc, iron and copper - but remember that there is NO reaction between dilute acid and copper!)

KEY POINTS:

- The reactivity series is a list of metals placed in order of their reactivity with oxygen, water and dilute hydrochloric or sulphuric acid.
- Metal + Oxygen ⟶ Metal Oxide. • Metal + Water ⟶ Metal Hydroxide + Hydrogen.
- Metal + Acid ⟶ Salt + Hydrogen with metal chlorides being formed when using hydrochloric acid and metal sulphates being formed when using sulphuric acid.

REACTIVITY SERIES II - Displacement reactions

A DISPLACEMENT REACTION is one in which a **MORE REACTIVE** metal displaces a **LESS REACTIVE** metal from a compound in a chemical reaction.

There's just one "incredibly important rule" to remember!

IF THE PURE METAL IS HIGHER IN THE REACTIVITY SERIES THAN THE METAL IN THE COMPOUND, THEN DISPLACEMENT WILL HAPPEN.

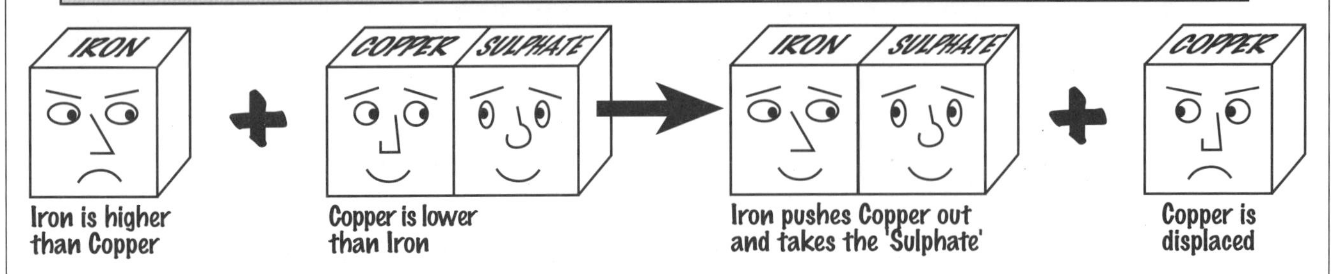

This is why an iron nail becomes coated with Copper when it's put in Copper Sulphate Solution, and why the blue solution becomes clear.

SOME MORE EXAMPLES OF DISPLACEMENT

Example No. 1

ZINC + COPPER SULPHATE SOLUTION

Remember the "Rule". Which is HIGHER in the Reactivity series?

ZINC + COPPER SULPHATE ⟶ ZINC SULPHATE + COPPER

Yes! Zinc is higher so it displaces the copper forming Zinc Sulphate.

Example No. 2

GOLD + COPPER SULPHATE SOLUTION

Remember the "Rule". Which is HIGHER in the Reactivity series?

GOLD + COPPER SULPHATE ⟶ COPPER SULPHATE + GOLD

No! Gold is lower in the series than copper so no reaction takes place.

Example No. 3

ALUMINIUM POWDER + POWDERED IRON OXIDE

Remember the "Rule". Which is HIGHER in the Reactivity series?

ALUMINIUM + IRON OXIDE ⟶ ALUMINIUM OXIDE + IRON

Yes! Aluminium is higher and as this gives out loads of heat, the iron is molten.

Summing Up

- Look at the Reactivity Series ...
- ... If the pure metal is higher than the metal in the metal compound ...
- ... then simply swop the metals around!!
- If it isn't there's no reaction.

KEY POINTS:

- A reaction where one metal displaces another metal from a compound is a displacement reaction.
- Displacement can only take place if the pure metal is higher in the reactivity series than the metal in the compound.
- If the pure metal is lower in the reactivity series then nothing happens and there is no displacement reaction.

EXTRACTION OF METALS

Most metals are extracted from their ORES, very often the OXIDES of the metals. To extract the metal, oxygen must be removed from the metal oxide. This is REDUCTION and the method of EXTRACTION depends on the position of the metal in the REACTIVITY SERIES.

METAL ELEMENT	EXTRACTION PROCESS
POTASSIUM	Metals **ABOVE ZINC** must be extracted by ELECTROLYSIS.
SODIUM	• ELECTROLYSIS is the breaking down of a compound ...
CALCIUM	• ... containing IONS ...
MAGNESIUM	• ... into its ELEMENTS ...
ALUMINIUM	• ... by using an ELECTRIC CURRENT.
*ZINC	Metals **BELOW ZINC** are REDUCED BY HEATING WITH CARBON.
IRON	• CARBON is above these metals in the reactivity series ...
COPPER	• ... so it is used as the REDUCING AGENT.
GOLD	NO EXTRACTION NEEDED as GOLD is so UNREACTIVE it EXISTS NATURALLY.

INCREASED REACTIVITY ↑

* POSITION OF CARBON (very important).

CORROSION OF METALS

Corrosion occurs when a metal reacts with OXYGEN from the air to form an OXIDE. This is OXIDATION. The rate of corrosion of a metal depends on its position in the REACTIVITY SERIES.

THE HIGHER THE POSITION OF THE METAL IN THE REACTIVITY SERIES ...
... THE QUICKER IT CORRODES.

(A) CORROSION OF IRON

The corrosion of IRON and STEEL is called RUSTING. It is a serious problem as they are the most commonly used metals.

For rusting to occur TWO things are necessary ... • OXYGEN ... and • WATER.

IRON (or steel) + OXYGEN ⟶ IRON OXIDE ...

... and then, ... IRON OXIDE + WATER ⟶ HYDRATED IRON OXIDE (Brown crumbly rust).

This can weaken the metal and reduce the useful life of many objects. It needs to be stopped.

PREVENTION

1. COATING THE SURFACE WITH A PROTECTIVE LAYER (to prevent contact with air or water)

- ... PAINT ...
- ... OIL or GREASE ...
- ... PLASTIC ...
- ... TIN or CHROMIUM PLATING.

e.g. Cars are painted!!

PROTECTIVE LAYER — IRON — NO RUSTING OCCURS

HOWEVER

SCRATCH! THEREFORE AIR AND WATER CAN REACH THE IRON — IRON — RUSTING OCCURS

2. COATING THE SURFACE WITH A MORE REACTIVE METAL ("SACRIFICIAL PROTECTION")

An ideal metal for this is ZINC (magnesium can also be used), since it is reasonably cheap, and is HIGHER UP IN THE REACTIVITY SERIES than IRON'. Because of this ...

- ... Oxygen will react with the Zinc rather than the Iron ...
- ... until ALL THE ZINC HAS COMPLETELY CORRODED AWAY. (The Zinc is 'sacrificed' to save the Iron)

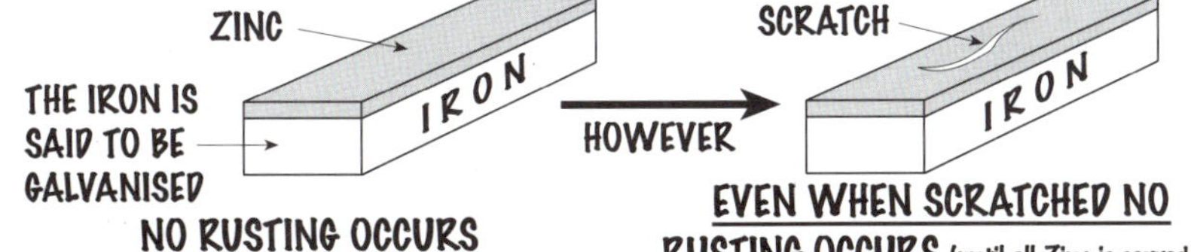

- For this reason, steel girders are GALVANISED with Zinc ...
- ... and Zinc blocks are attached to the metal hull of ships

(B) CORROSION OF ALUMINIUM

Aluminium also corrodes by reacting with oxygen, however the ALUMINIUM OXIDE formed prevents any further corrosion as it covers the surface of the metal.

KEY POINTS:

- Corrosion occurs when a metal reacts with oxygen to form an oxide.
- Corrosion of iron can be prevented by coating its surface with a protective layer or a more reactive metal.

ACIDS AND BASES

pH SCALE

This tells us the ACIDITY, ALKALINITY, or NEUTRALITY of a solution.

VERY ACIDIC ⟷ SLIGHTLY ACIDIC						NEUTRAL	SLIGHTLY ALKALINE ⟷ VERY ALKALINE						
1	2	3	4	5	6	7	8	9	10	11	12	13	14

When used with UNIVERSAL INDICATOR, we get the following range of colours:-

RED	ORANGE	YELLOW	GREEN	BLUE	NAVY BLUE	PURPLE

eg. Battery Acid; Lemon Juice, Vinegar; Soda Water; Water; Soap; Baking Powder; Washing Soda; Oven Cleaner; Potassium Hydroxide

REACTIONS OF DILUTE HYDROCHLORIC ACID AND DILUTE SULPHURIC ACID

	DILUTE HYDROCHLORIC ACID	DILUTE SULPHURIC ACID
METALS	• Forms METAL CHLORIDES and HYDROGEN (see S.C.5)	• Forms METAL SULPHATES and HYDROGEN (see S.C.5)
METAL OXIDES AND HYDROXIDES	• Forms METAL CHLORIDES and WATER e.g. Sodium oxide + Hydrochloric acid ⟶ Sodium hydroxide + Sulphuric acid ⟶	• Forms METAL SULPHATES and WATER Sodium chloride + water Sodium sulphate + water
METAL CARBONATES AND HYDROGEN CARBONATES	• Forms METAL CHLORIDES and WATER and CARBON DIOXIDE e.g. Calcium carbonate + Hydrochloric acid ⟶ Calcium hydrogen carbonate + Sulphuric acid ⟶	• Forms METAL SULPHATES and WATER and CARBON DIOXIDE Calcium chloride + water + carbon dioxide Calcium sulphate + water + carbon dioxide
AMMONIA	• Forms AMMONIUM CHLORIDE e.g. Ammonia + Hydrochloric acid ⟶ Ammonium chloride	• Forms AMMONIUM SULPHATE e.g. Ammonia + Sulphuric acid ⟶ Ammonium sulphate

NEUTRALISATION

This a reaction between an ACID and an ALKALI which gives us a NEUTRAL (pH of 7) solution.

Everyday examples ...
- INDIGESTION occurs when there is too much acid in the stomach. Can be neutralised by taking indigestion tablets which contain an ALKALI.
- ACIDIC SOILS can be treated with LIME, an alkali, to cancel out their acidity. Acidic soils result in poor crop growth.

Neutralisation - the process

TWO important facts ...
- ACIDS are substances that form hydrogen IONS, $H^+_{(aq)}$, when added to water.
- SOLUBLE BASES or ALKALIS are substances that form hydroxide IONS, $OH^-_{(aq)}$, when added to water.

  When neutralisation occurs $H^+_{(aq)}$ ions from the acid and $OH^-_{(aq)}$ ions from the alkali JOIN TOGETHER.

$H^+_{(aq)} + OH^-_{(aq)} \longrightarrow H_2O_{(l)}$ i.e. water (pH of 7)

e.g. Hydrochloric acid + sodium hydroxide ⟶ sodium chloride + water

In terms of IONS: $H^+_{(aq)}Cl^-_{(aq)} + Na^+_{(aq)}OH^-_{(aq)} \longrightarrow Na^+_{(aq)}Cl^-_{(aq)} + H_2O_{(l)}$

THIS HAPPENS IN ALL NEUTRALISATION REACTIONS. $H^+_{(aq)} + OH^-_{(aq)} \longrightarrow H_2O_{(l)}$

NO LONGER ANY H^+ OR OH^- IONS PRESENT

KEY POINTS:

- The pH scale tells us the acidity, alkalinity or neutrality of a solution.
- The reaction between an acid and an alkali which gives a neutral solution is called Neutralisation.

RATES OF REACTION I

CHEMICAL REACTIONS occur when the particles of substances collide with each other. The substances at the start of the reaction are called the **REACTANTS** while those left at the end are the **PRODUCTS**.
The **SPEED** or **RATE** of the reaction can be **INCREASED** in **FOUR** ways.

1. INCREASING THE TEMPERATURE OF THE REACTANTS

Low Temp

REACTANT (LIQUID)

REACTANT (SOLID)

High Temp

LIQUID PARTICLES MOVE FASTER

- The **HIGHER THE TEMPERATURE** of the **REACTANTS**, the **GREATER** the **FREQUENCY** and **ENERGY** of **COLLISION** ...
- ... resulting in **INCREASED RATE OF REACTION** (i.e. conversion of **REACTANTS** to **PRODUCTS** is **QUICKER**)

2. INCREASING THE CONCENTRATION OF DISSOLVED REACTANTS

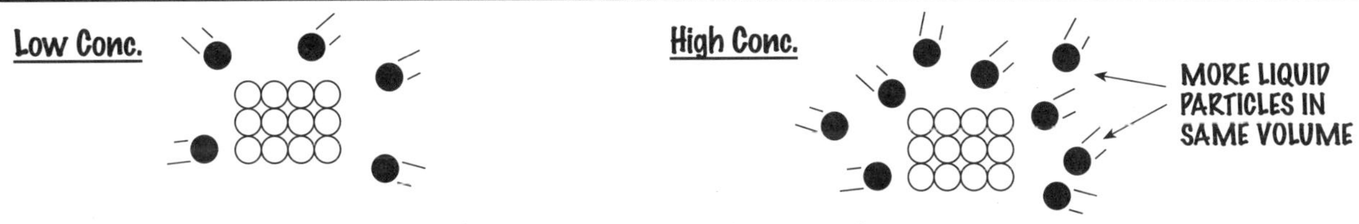

- The **GREATER THE CONCENTRATION** (i.e. more particles in the solution) ...
- ... the **GREATER THE FREQUENCY OF COLLISIONS** ... resulting in **INCREASED RATE OF REACTION**.

> • **IN THE CASE OF GASES <u>CONCENTRATION</u> AND <u>PRESSURE</u> MEAN THE SAME THING, SINCE IF A GAS IS AT HIGH PRESSURE ITS MOLECULES ARE CLOSE TOGETHER. i.e. ITS CONCENTRATION IS HIGH!**

3. INCREASING THE SURFACE AREA OF SOLID REACTANTS

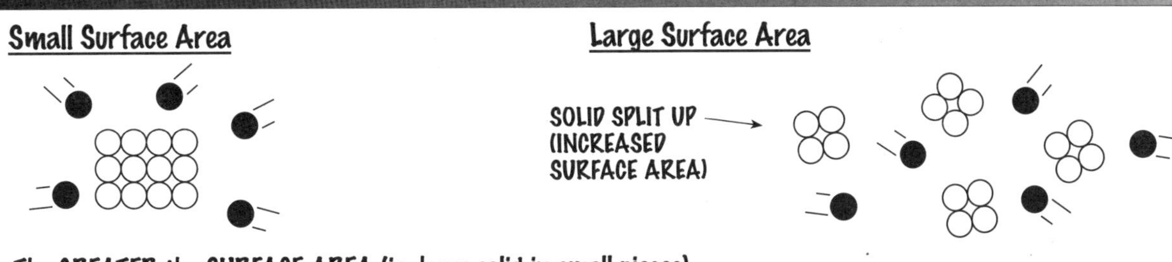

- The **GREATER** the **SURFACE AREA** (ie have solid in small pieces) ...
- the **GREATER** the **FREQUENCY** of **COLLISIONS** ... resulting in **INCREASED RATE OF REACTION**.

4. USING A CATALYST

A catalyst
- ... is a substance which **INCREASES THE RATE OF A CHEMICAL REACTION** ...
- ... **WITHOUT BEING USED UP** in the process.
- Catalysts are **VERY SPECIFIC** and so **DIFFERENT REACTIONS NEED DIFFERENT CATALYSTS**.

Examples are:- 1) the 'cracking' of hydrocarbons using broken pottery.
2) the manufacture of Ammonia (**HABER PROCESS**) using **IRON**.

When used on a large scale industrial costs are dramatically reduced, because of increased **YIELD**.

KEY POINTS:

- The speed or rate of a chemical reaction can be increased by increasing the temperature of the reactants, increasing the concentration of the dissolved reactants, increasing the surface area of solid reactants and by using a catalyst.

ANALYSING RATE OF REACTION

A chemical reaction can be analysed by measuring:

- HOW FAST THE PRODUCTS ARE FORMED ... or • HOW FAST THE REACTANTS ARE USED UP.

Below is a typical graph of the SAME REACTION performed under TWO DIFFERENT CONDITIONS.

SOME VERY IMPORTANT FACTS ABOUT THE GRAPH.

- The SLOPE of the graph is GREATEST AT THE BEGINNING ...
  ... when there are MORE REACTANTS to collide with each other.
- It then steadily DECREASES as the reactants are used up ...
  ... until the end of the reaction.

REACTION 'A' WAS COMPLETED FASTER THAN REACTION 'B' ...
... due to one of the following factors:-

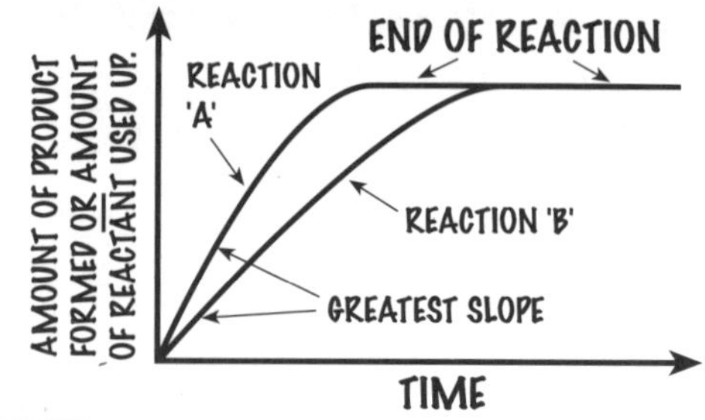

(1) TEMPERATURE OF REACTANTS in 'A' > 'B' (2) CONCENTRATION OF REACTANTS IN 'A' > 'B'

(3) SURFACE AREA OF REACTANTS IN 'A' > 'B' (4) USE OF A CATALYST (Reaction 'A' had catalyst, 'B' did not).

THREE EXAMPLES

1. Reaction between CALCIUM CARBONATE and dilute HYDROCHLORIC ACID

The rate of reaction can be analysed by ...

- MEASURING THE VOLUME OF CARBON DIOXIDE GIVEN OFF.

The equation: Calcium carbonate + hydrochloric acid ⟶ Calcium chloride + carbon dioxide + water

$$CaCO_{3(s)} + 2HCl_{(aq)} \longrightarrow CaCl_{2(aq)} + CO_{2(g)} + H_2O_{(l)}$$

Rate of reaction can be altered by changing ...

- Surface Area of calcium carbonate ⟶ • INCREASED AREA, INCREASED RATE OF REACTION
- Concentration of hydrochloric acid ⟶ • INCREASED CONC., INCREASED RATE OF REACTION
- Temperature of hydrochloric acid ⟶ • INCREASED TEMP., INCREASED RATE OF REACTION

2. Reaction between SODIUM THIOSULPHATE solution and dilute HYDROCHLORIC ACID

The rate of reaction can be analysed by ...

- MEASURING THE AMOUNT OF SULPHUR WHICH IS PRECIPITATED.

The equation: Sodium thiosulphate + hydrochloric acid ⟶ sodium chloride + sulphur + sulphur dioxide + water

$$Na_2S_2O_{3(aq)} + 2HCl_{(aq)} \longrightarrow 2NaCl_{(aq)} + S_{(s)} + SO_{2(g)} + H_2O_{(l)}$$

Rate of reaction can be altered by changing ...

- Concentration of sodium thiosulphate ... ⟶ • INCREASED CONC., INCREASED RATE OF REACTION
- ... or conc. of hydrochloric acid.
- Temperature of sodium thiosulphate ... ⟶ • INCREASED TEMP., INCREASED RATE OF REACTION
- ... or temp of hydrochloric acid.

3. Decomposition of HYDROGEN PEROXIDE solution

The rate of reaction can be analysed by ...

- MEASURING THE VOLUME OF OXYGEN GIVEN OFF.
- A simple TEST for the gas given off in this reaction is to place a glowing wooden splint into the gas. If the splint relights, the gas is OXYGEN.

The equation: Hydrogen peroxide ⟶ water + oxygen

$$2H_2O_{2\,(aq)} \longrightarrow 2H_2O_{(l)} + O_{2\,(g)}$$

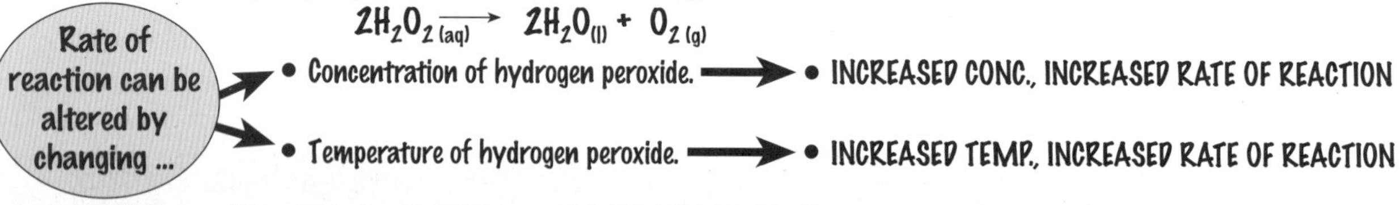

... and by ... ⟶ • Use of a CATALYST e.g. MANGANESE OXIDE ⟶ • INCREASED RATE OF REACTION

KEY POINTS:

- The rate of a chemical reaction can be analysed by measuring how fast the products are formed or how fast the reactants are used up.

USEFUL PRODUCTS FROM OIL I - Crude Oil

HOW OIL AND GAS WERE FORMED

- Formed over millions of years from dead **ORGANISMS**, mainly **PLANKTON** (tiny sea creatures) ...
- ... which fell to the ocean floor and were covered by **MUD SEDIMENTS**. It is a **FOSSIL FUEL**.
- Action of **HEAT** and **PRESSURE** in the **ABSENCE OF OXYGEN** ...
- ... caused the production of **CRUDE OIL** and **NATURAL GAS**, which becomes trapped under **NON-POROUS** layers of sediment.
- **Both are NON-RENEWABLE ENERGY RESOURCES with LIMITED SUPPLIES LEFT!!**

WHAT CRUDE OIL IS

- **CRUDE OIL** is a **MIXTURE** of compounds most of which ...
- ... are made up only of **CARBON** and **HYDROGEN** atoms called **HYDROCARBONS**.
- Some of these **HYDROCARBONS** have very short chains of **CARBON** atoms (less than 20) ...
- ... Some of these **HYDROCARBONS** have very long chains of **CARBON** atoms (more than 20).

HOW WE SEPARATE THIS MIXTURE

Crude oil on its own isn't a great deal of use. (Ged Clampitt didn't know what to do with it!) We need to separate it into its different **FRACTIONS** all of which have their own **PARTICULAR CHARACTERISTICS**.

For instance, **the LARGER the HYDROCARBON** molecules are in a **FRACTION**;

- **the LESS EASILY it FLOWS**
- **the LESS EASILY it IGNITES**
- **the LESS VOLATILE it is** (doesn't vaporise as easily)
- **the HIGHER its BOILING POINT is.**

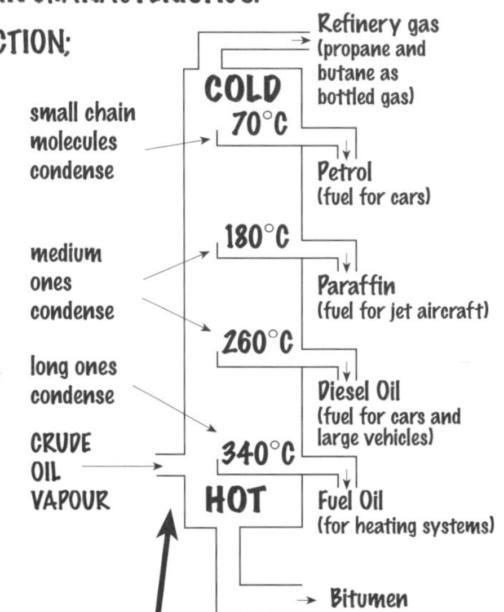

This last feature makes it possible for us to separate the **FRACTIONS** by

- **EVAPORATING** the oil by heating ...
- ... then allowing it to **CONDENSE** ...
- ... at a **RANGE of DIFFERENT TEMPERATURES** ...
- ... when it will form **FRACTIONS** each of which ...
- ... will contain molecules with a **SIMILAR NUMBER OF CARBON ATOMS.**
- **THIS IS CALLED** fractional distillation, and is done in a **FRACTIONATING COLUMN.**

These molecules are called **ALKANES**, the simplest being **METHANE**, CH_4.

CRACKING

Because the **SHORTER CHAIN HYDROCARBONS** release energy more quickly by **BURNING**, there is a greater demand for them.

- Therefore **LONGER CHAIN HYDROCARBONS** are **'CRACKED'** into **SMALLER ONES** ...
- ... by passing their **VAPOUR** over a **CATALYST** at **HIGH TEMPERATURE** and **PRESSURE**.
- This produces **ALKENES**, the simplest being **ETHENE**, C_2H_4.

KEY POINTS:

- Oil and gas are non-renewable energy resources.
- Crude oil is a mixture of compounds most of which are hydrocarbons.
- Crude oil is separated into its fractions using a fractionating column.
- The molecules produced are called alkanes.
- Cracking is a method used to break down long chain hydrocarbons into smaller ones with the production of alkenes.

ALKANES - Saturated Hydrocarbons

- HYDROCARBONS are molecules which contain only HYDROGEN and CARBON.
- Carbon atoms can form FOUR SINGLE COVALENT BONDS with other atoms.
- The 'SPINE' of a HYDROCARBON is made up of a chain of CARBON ATOMS.
- When each **CARBON ATOM** in the 'spine' forms **FOUR SINGLE COVALENT BONDS** ...
- ... we say the **HYDROCARBON is SATURATED** and we call it an **ALKANE.**

Examples of Alkanes:

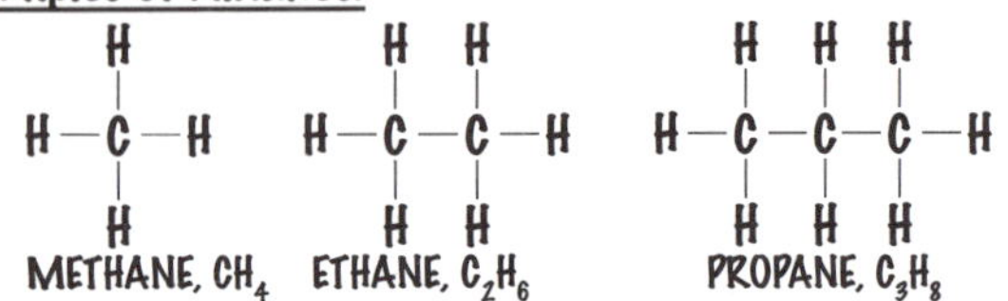

- All the CARBON atoms are ...
- ... linked to 4 other atoms.
- They are "FULLY OCCUPIED" ...
- ... or SATURATED.
- All the BONDS are SINGLE BONDS.

Because all their bonds are 'occupied' they are fairly UNREACTIVE, although they do burn well.

ALKENES - Unsaturated Hydrocarbons

- These are the products of the cracking of alkanes where ...
- ... carbon atoms can also form DOUBLE COVALENT BONDS with other atoms
- In this case they would join with less than 4 other atoms.
- When the **CARBON atoms** in the 'spine' of a hydrocarbon have ...
- ... at least **ONE DOUBLE COVALENT BOND,** ...
- ... we say the **HYDROCARBON is UNSATURATED** and we call it an **ALKENE.**

TEST FOR ALKANE OR ALKENE
BROMINE WATER reacts with ALKENES changing colour from ORANGE to CLEAR. NO REACTION with ALKANES

Examples of Alkenes:

Look at the DOUBLE BONDS (=)

ETHENE, C_2H_4 PROPENE, C_3H_6

- Not all the CARBON atoms are ...
- ... linked to 4 other atoms.
- They are NOT all "FULLY OCCUPIED" ...
- ... ie, they are UNSATURATED.
- A DOUBLE BOND is present.

Because of this DOUBLE BOND the ALKENES have the potential to join other atoms and are REACTIVE. This makes them useful for making other molecules, especially POLYMERS.

POLYMERISATION

- The small molecules (e.g. ethene) can be described as MONOMERS.
- When lots of MONOMERS join together they form a POLYMER. This is POLYMERISATION.
- e.g. POLY(ETHENE) which is used to make PLASTIC BAGS and BOTTLES is ...
- ... formed from ETHENE molecules joined together.

ADDITION POLYMERISATION

- Because ALKENES are UNSATURATED, they are very good at joining together and ...
- ... when they do so without producing another substance, we call this ...
- ... **ADDITION POLYMERISATION.**

An example of Addition Polymerisation - Poly(ethene)

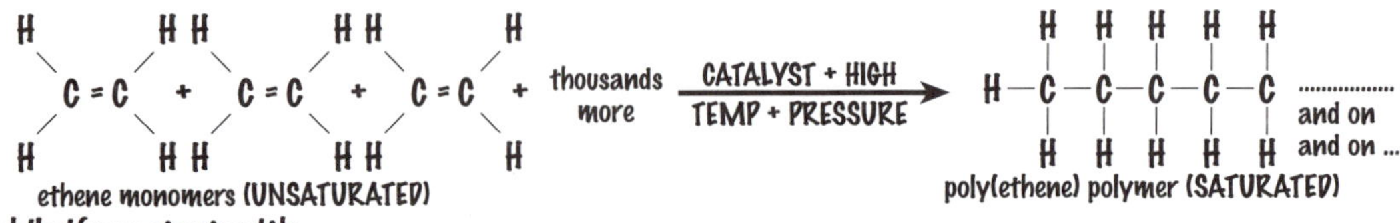

... while if we start with ...

PROPENE MONOMERS → CATALYST + HIGH TEMP + PRESSURE → **POLY(PROPENE) POLYMER** (Used for CRATES and ROPES)

KEY POINTS:

- Alkanes are saturated hydrocarbons where all the bonds are single bonds.
- Alkenes are unsaturated hydrocarbons where at least one double bond is present.
- The joining together of alkenes without the production of another substance is called addition polymerisation.

USEFUL PRODUCTS FROM OIL III - Burning Hydrocarbons

As we have seen CRUDE OIL is a very important MATERIAL as it provides FUEL and PLASTICS as well as hundreds of other products that we use everyday. However there is a PRICE TO PAY for all of this!, although it is only in recent years that there has been an awareness of the problems being created.

PROBLEMS CAUSED BY BURNING HYDROCARBONS

- When a substance burns it reacts with OXYGEN ...
- ... to form OXIDES.
- When a HYDROCARBON burns ...
- ... WASTE PRODUCTS are formed which are released into the ATMOSPHERE.

THESE ARE EXOTHERMIC REACTIONS AS PLENTY OF HEAT IS GIVEN OUT.

Example

METHANE + OXYGEN ⟶ CARBON DIOXIDE + WATER

However if the supply of OXYGEN is limited ...

METHANE + OXYGEN ⟶ CARBON MONOXIDE + WATER

... and if the supply of OXYGEN is very limited ...

METHANE + OXYGEN ⟶ CARBON + WATER

Under normal combustion the waste products CARBON DIOXIDE and WATER are identified by these tests:

Carbon dioxide ...	• When CO_2 is passed through CALCIUM HYDROXIDE SOLUTION (LIMEWATER) ... • ... the solution turns from COLOURLESS to MILKY due to formation of CALCIUM CARBONATE precipitate.
Water ...	• Test it for it's BOILING POINT ... should be 100°C. • TURNS ANHYDROUS COPPER SULPHATE from WHITE TO BLUE. • TURNS ANHYDROUS COBALT CHLORIDE from BLUE TO PINK.

GLOBAL WARMING - Effect of increased carbon dioxide

1. Increased production of CARBON DIOXIDE - GLOBAL WARMING

- CARBON DIOXIDE is one of the GREENHOUSE GASES
- INCREASED BURNING due to INDUSTRIAL REVOLUTION ...
- ... has caused INCREASED RELEASE of CARBON DIOXIDE ...
- ... from FOSSIL FUELS.

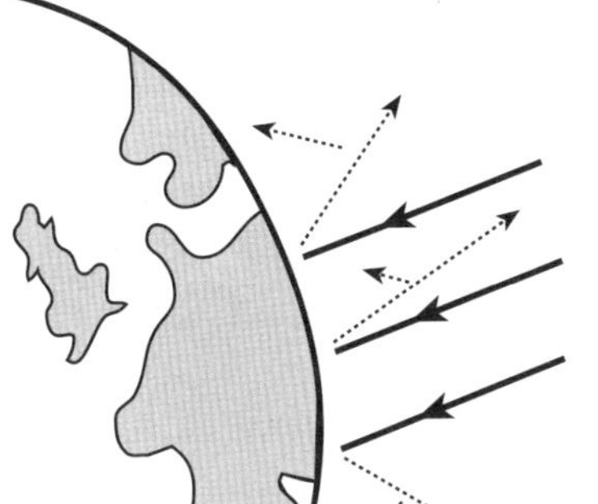

- Light from the sun reaches the earth ...
- ... and passes through the atmosphere.
- This WARMS up the planet which ...
- ... then radiates this heat energy ...
- ... back into SPACE.
- CARBON DIOXIDE helps to trap some of this energy ...
- ... which helps to keep the planet WARM.
- Too much CO_2 however, leads to ...
- ... too much heat being retained ...
- i.e., GLOBAL WARMING.

2. Production of CARBON MONOXIDE

- MOST CARBON MONOXIDE emissions are from the exhausts of motor vehicles.
- It is a COLOURLESS, ODOURLESS and POISONOUS gas as ...
- ... it combines with HAEMOGLOBIN, found in red blood cells, ...
- ... preventing the haemoglobin carrying OXYGEN around the body.
- This SHORTAGE of OXYGEN leads to DIZZINESS, then UNCONSCIOUSNESS, and ultimately DEATH!!

KEY POINTS:

- The combustion of a hydrocarbon releases waste products which are released into the atmosphere.
- Increased production of carbon dioxide is responsible for global warming.
- Carbon monoxide is produced when hydrocarbons are burned in poor ventilation.

OIL SPILLAGES

SPILLAGES from OIL TANKERS at SEA are ... → ... due to → ... COLLISION, OVER-TURNING, ACCIDENTAL or ILLEGAL DUMPING which ... → ... causes → ... UNTOLD DAMAGE to MARINE ANIMALS even DEATH ... → ... while → ... if the spill is near the SHORE then BEACHES are BADLY POLLUTED by the incoming oil.

DISPOSAL OF POLYMERS

The trouble with MANUFACTURED POLYMERS (poly(ethene) etc) is that they are NON-BIODEGRADABLE ... → ... which means ... → ... that MICRO-ORGANISMS are unable to break them down into simpler substances. → However ... → ... RECYCLING is an option but unfortunately one plastic looks very much like another one when it comes to sorting them out!

STRUCTURE AND CHANGES SUMMARY QUESTIONS

1. What are atoms made up of?
2. What is the ATOMIC NUMBER of an element?
3. How are the elements of the periodic table arranged?
4. What can we say about elements of the same group?
5. Name the SIX NOBLE GASES.
6. List the characteristics of the Noble gases.
7. Explain the significance of a full outer electron shell.
8. List the main characteristics of the alkali metals.
9. Give balanced equations to illustrate their reaction with a) water b) oxygen and c) halogens.
10. What are the products of the electrolysis of sodium chloride solution?
11. List the main characteristics of the halogens.
12. Give balanced equations to illustrate their reaction with a) alkali metals and b) hydrogen.
13. How does ATOMIC NUMBER affect the properties of alkali metals and halogens?
14. Name the 3 most reactive elements in the REACTIVITY SERIES.
15. On which four reactions is the series based?
16. What is produced when a METAL reacts with a) oxygen b) water c) dilute Hydrochloric acid and d) dilute Sulphuric acid?
17. What is the 'incredibly important rule' for DISPLACEMENT REACTIONS?
18. Complete the equation: Magnesium + Zinc Sulphate. ⟶
19. Complete the equation: Zinc + Iron Sulphate. ⟶
20. Complete the equation: Copper + Iron Oxide. ⟶
21. How are metals ABOVE ZINC extracted from their ores?
22. How are metals BELOW ZINC extracted from their ores?
23. Why does GOLD exist naturally?
24. How does the position of a metal in the reactivity series affect how quickly it corrodes?
25. Write down a WORD EQUATION for rusting.
26. Describe how a PROTECTIVE LAYER can be used to prevent rusting.
27. Describe how a MORE REACTIVE METAL can be used to prevent rusting.
28. What pH is NEUTRAL? What is the most ACIDIC pH?
29. Write down WORD EQUATIONS for the reaction of DILUTE HYDROCHLORIC ACID and DILUTE SULPHURIC ACID with a) metals b) metal oxides c) metal carbonates and d) ammonia.
30. Write down the equation for NEUTRALISATION.
31. Describe how and why a) increasing temperature b) increasing concentration and c) increasing surface area increases the rate of reaction.
32. Describe how and why catalysts are used particularly in industrial processes.
33. How can a chemical reaction be analysed? Sketch a graph to show a 'slow' reaction and a 'faster' reaction.
34. How is OIL split into fractions?
35. Explain what we mean by 'CRACKING'.
36. What is a) an ALKANE b) an ALKENE and c) a POLYMER?
37. What is ADDITION POLYMERISATION?
38. What are the problems associated with the BURNING OF HYDROCARBONS?
39. What are the problems associated with OIL SPILLAGES?
40. What are the problems associated with the DISPOSAL OF POLYMERS?

FORCE AND TRANSFERS

ELECTRICAL CIRCUITS AND MAINS ELECTRICITY 1 FT 1

CURRENT AND VOLTAGE - some basics

An ELECTRIC CURRENT is a FLOW of ...
... • ELECTRONS in metals.
or • IONS in certain substances when dissolved in water or melted.
A current will only flow if there is a VOLTAGE across the ends of the circuit or component.
A cell, battery or power pack usually provides this for us in the lab.

MEASURING CURRENT AND VOLTAGE

An electrical circuit is the means of transferring electrical energy from batteries and other sources to other components.
This electrical energy is measured in Joules.

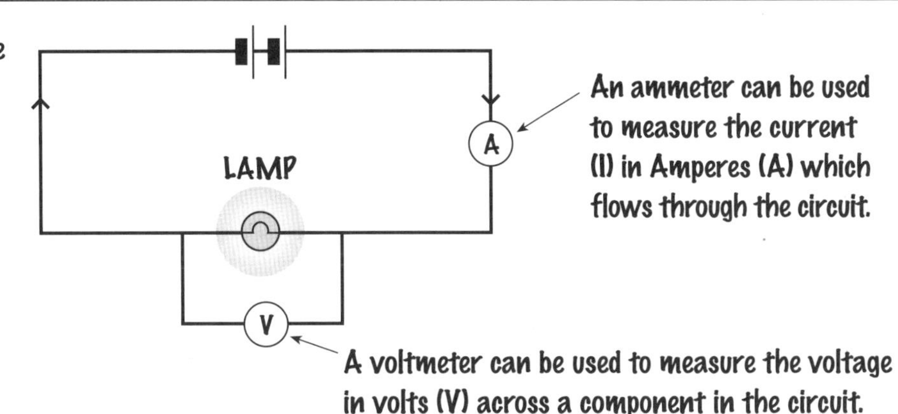

An ammeter can be used to measure the current (I) in Amperes (A) which flows through the circuit.

A voltmeter can be used to measure the voltage in volts (V) across a component in the circuit.

SYMBOLS USED IN CIRCUITS

The following standard symbols should be known. You may be asked to interpret and/or draw circuits using the following standard symbols.

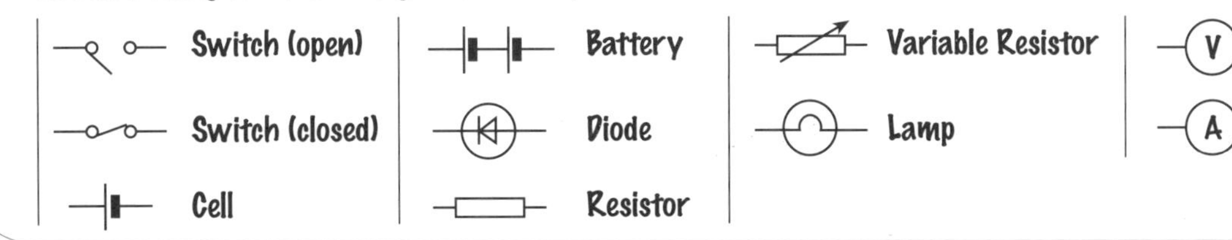

HEATING EFFECT IN RESISTORS

Very simply ... • MOVING ELECTRONS collide with ATOMS within a RESISTOR ...
• ... GIVING UP THEIR ENERGY which results ...
• ... in the TEMPERATURE of the RESISTOR INCREASING.

Three common electrical appliances which use this effect are:

1. Hairdrier 2. Immersion heater 3. Light bulb

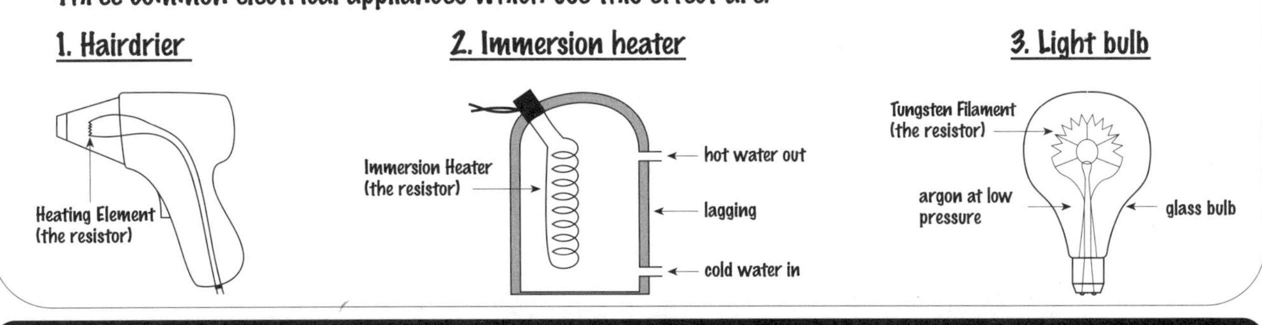

KEY POINTS:

- An electric current is a flow of electrons in metals or ions in certain substances when dissolved in water or melted.
- Current is measured using an ammeter and voltage is measured using a voltmeter.

Very often circuits contain TWO or more COMPONENTS, though not necessarily the same ones! Here we will consider the TWO types of circuit possible ...

- ... SERIES and ...
- ... PARALLEL ...

... together with the ADVANTAGES and DISADVANTAGES of having LIGHT BULBS wired into each type of circuit.

SERIES CIRCUITS

1. The SAME CURRENT flows through ...
   ... EACH COMPONENT in the circuit.

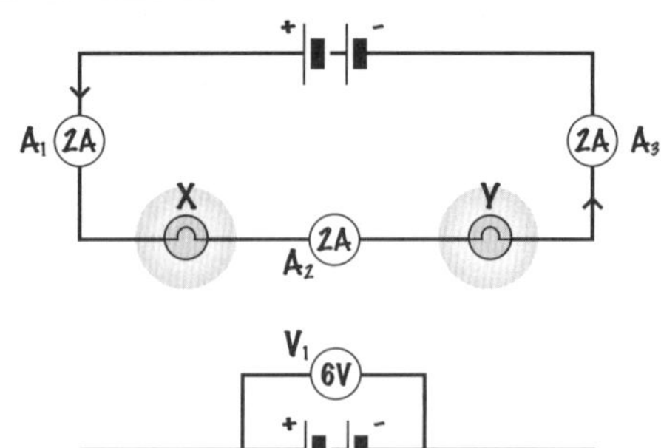

$$A_1 = A_2 = A_3$$

2. The TOTAL VOLTAGE is SHARED ...
   ... BETWEEN THE COMPONENTS in the circuit.

$$V_1 = V_2 + V_3$$

In this case the two bulbs have identical resistances and therefore the voltage is split equally but the voltage could be split 4 and 2 or 5 and 1 (for example) if bulbs of different resistance were used.

For light bulbs in series:

ADVANTAGES

- Supply voltage is divided between the bulbs so its less dangerous if it is the mains supply of 240V!!
- Less 'drain' on the supply voltage as current supplied is LESS than in a parallel circuit.

DISADVANTAGES

- Can't have one bulb lit on its own, ALL ON or ALL OFF!
- If one bulb fails, they ALL fail.

PARALLEL CIRCUITS

1. The TOTAL CURRENT in the CIRCUIT ...
   ... is the SUM of the CURRENTS THROUGH EACH COMPONENT.

$$A_1 = A_2 + A_3 = A_4$$

2. The CURRENT through each component depends on its RESISTANCE. The greater the resistance of a component, the smaller the current.
3. There is the SAME VOLTAGE across each component $V_1 = V_2 = V_3$

For light bulbs in parallel:

ADVANTAGES

- If one bulb fails, then other bulbs are still working.
- Possible to have one bulb lit on its own.

DISADVANTAGES

- All connected directly to supply voltage so it can be dangerous if it is the mains supply of 240V!!
- More 'drain' on the supply voltage as current supplied is MORE than in a series circuit.

KEY POINTS:

- In a series circuit the same current flows through each component in the circuit and the total voltage is shared between the components in the circuit.
- In a parallel circuit the total current in the circuit is the sum of the currents through each component, the current passing through each component depends on its resistance, and there is the same voltage across each component.

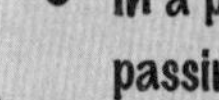

The CURRENT in an electrical circuit will CHANGE if there is a ...

- ... CHANGE in the VOLTAGE within the circuit and/or a ...
- ... CHANGE in the RESISTANCE within the circuit ...

... where resistance is a measure of how hard it is to get the current through a conductor or component at a particular voltage. It's unit is OHMS (Ω).

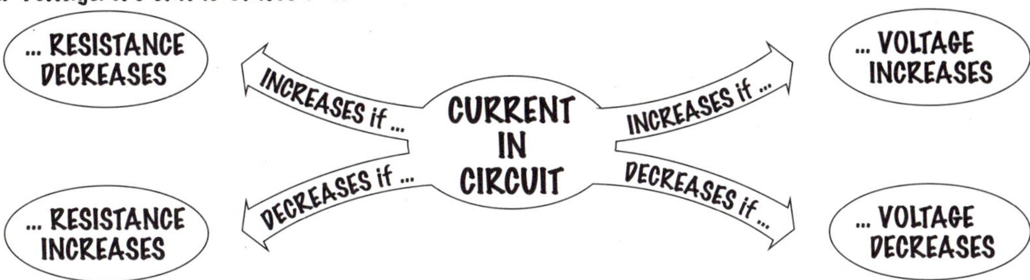

The relationship between voltage, current and resistance is:

VOLTAGE (Volts) = CURRENT (Amps) x RESISTANCE (Ohms)

$V = {}^{*}I \times R$

* (we use the letter I for current)

V / I x R

THIS FORMULA IS NOT GIVEN - SO LEARN IT.

EXAMPLE

A 6 volt battery is connected to a single lamp and a current of 2 Amps passes through the lamp. What is the resistance of the lamp?

Using our equation, (which is rearranged) $\text{Resistance} = \frac{\text{Voltage}}{\text{Current}}$, $R = \frac{6V}{2A}$, $R = 3$ Ohms (Ω)

CURRENT - VOLTAGE GRAPHS

These show us how the current through a component varies with the voltage across it.

① RESISTOR AT CONSTANT TEMP.

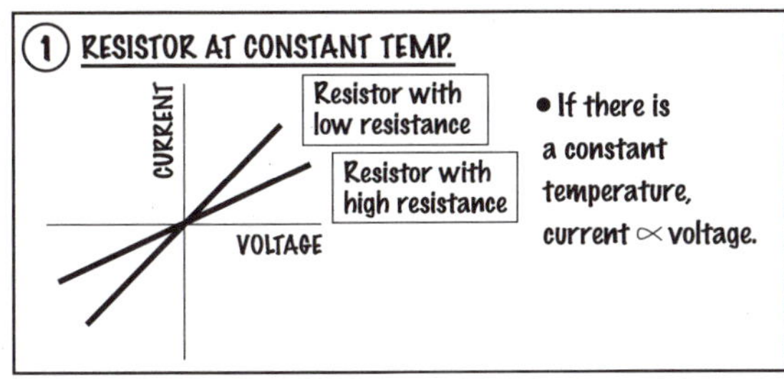

- If there is a constant temperature, current ∝ voltage.

② WIRES MADE OF DIFFERENT METALS AT CONST. TEMP.

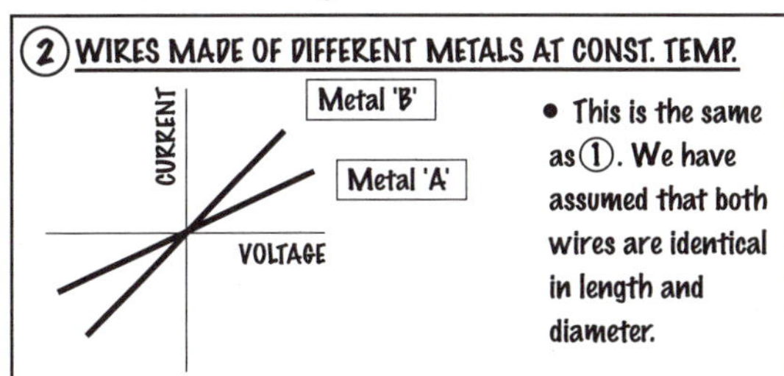

- This is the same as ①. We have assumed that both wires are identical in length and diameter.

③ A FILAMENT BULB

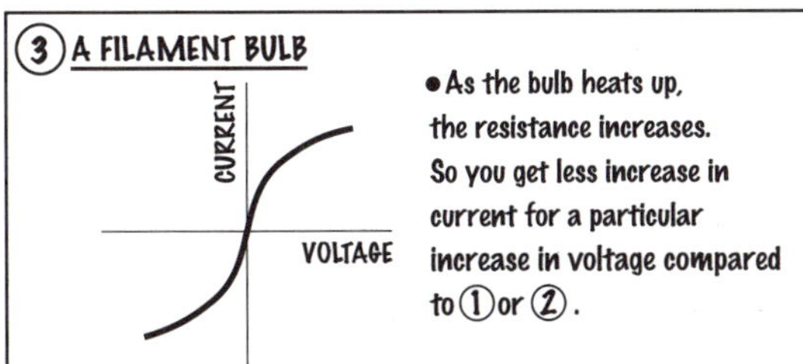

- As the bulb heats up, the resistance increases. So you get less increase in current for a particular increase in voltage compared to ① or ②.

④ A DIODE

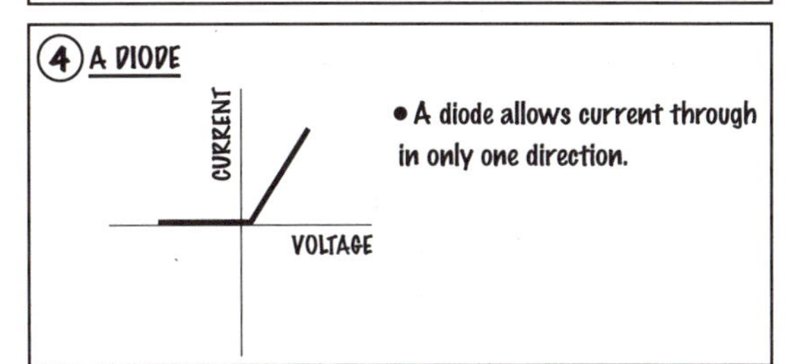

- A diode allows current through in only one direction.

⑤ LIGHT DEPENDENT RESISTORS (L.D.R)

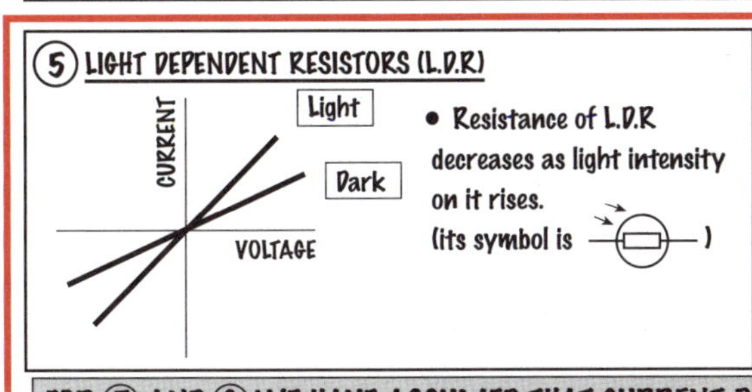

- Resistance of L.D.R decreases as light intensity on it rises. (its symbol is)

⑥ THERMISTOR

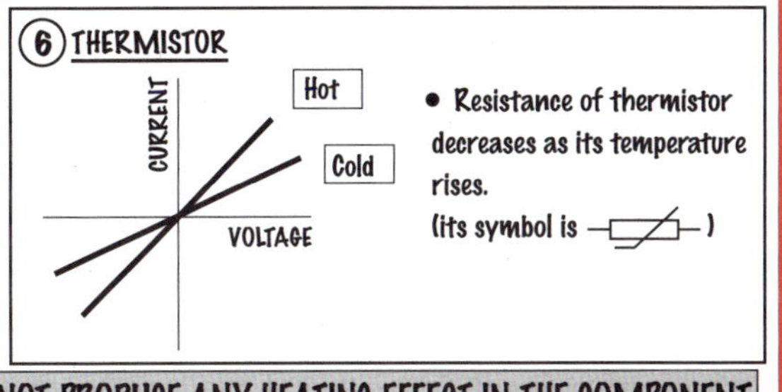

- Resistance of thermistor decreases as its temperature rises. (its symbol is)

FOR ⑤ AND ⑥ WE HAVE ASSUMED THAT CURRENT DOES NOT PRODUCE ANY HEATING EFFECT IN THE COMPONENT

KEY POINTS:

- Voltage (V) = Current (A) x Resistance (Ω)
- Current-Voltage graphs can be used to describe the resistance of a component.

POWER

When an ELECTRIC CURRENT flows through a circuit ...

- ... ENERGY is TRANSFERRED from the BATTERY or SUPPLY VOLTAGE ...
- ... to the COMPONENTS in that circuit.

The **RATE** of this **ENERGY TRANSFER** is the **POWER** in Joules/sec or **WATTS**

The relationship: **ELECTRICAL POWER (Watts) = VOLTAGE (Volts) x CURRENT (Amps)**

FUSES

Very simply ...

- A FUSE is a SHORT, THIN piece of WIRE ...
- ... with a LOW MELTING POINT.
- When the CURRENT passing through it EXCEEDS ...
- ... the CURRENT RATING of the fuse, ...
- ... the fuse wire gets HOT and BURNS OUT or BREAKS.
- This PREVENTS DAMAGE to CABLE or APPLIANCE through the possibility of OVERHEATING.

3 AMP 13 AMP

TOO LARGE A CURRENT → GREATER THAN CURRENT RATING OF FUSE → FUSE BURNS OUT → CIRCUIT IS BROKEN → NO CURRENT FLOWS → CABLE OR APPLIANCE IS PROTECTED

EXAMPLES

1. A domestic iron has a rating plate stuck on it as shown below. Calculate the current that passes through the iron when in use and the current rating of the fuse which should be fitted into its plug if a 3A, 5A, 10A and 13A fuse are available.

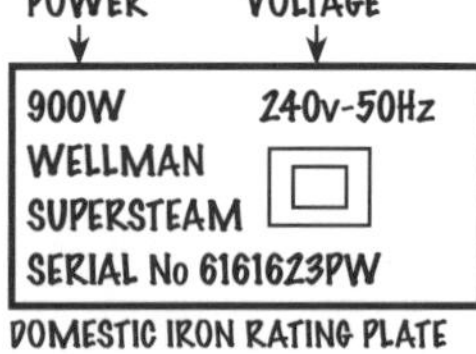

DOMESTIC IRON RATING PLATE

THE RATING PLATE GIVES US THE <u>POWER</u> AND <u>VOLTAGE</u> and so ...

Using our equation above: Current $= \frac{\text{Power}}{\text{Voltage}}$, $I = \frac{900W}{240V}$, $I =$ <u>3.75 Amps</u>
(which is rearranged)

Fuse chosen for plug should be <u>5A</u> because its just above the normal current flowing while ...

- A 3A fuse if chosen would burn out immediately (fairly obvious) and ...
- ... a 10A or 13A fuse would work <u>BUT</u> both would allow a large and potentially dangerous current to flow before burn out occurred.

2. A tablelamp connected to the mains voltage at 240V contains a filament of resistance 960Ω.
What current rating should the fuse in its plug have?
THIS TIME WE ARE GIVEN THE <u>VOLTAGE</u> AND <u>RESISTANCE</u> and so ...
Using the equation from FT.3: Current $= \frac{\text{Voltage}}{\text{Resistance}}$, $I = \frac{240V}{960\Omega}$, $I =$ <u>0.25A</u>
(which is rearranged)
Fuse chosen for plug should be <u>3A</u>.

KEY POINTS:

- Electrical Power (W) = Voltage (V) x Current (A)
- A fuse is a safety device used to protect a cable or an appliance.

THREE PIN FUSED PLUG

N.B. No bare wires showing around screws!!

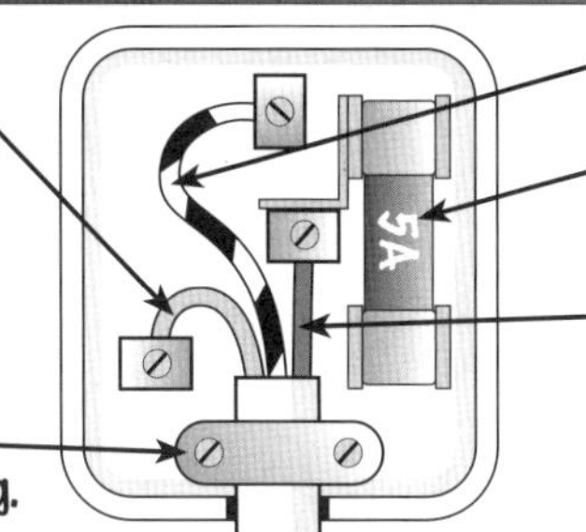

NEUTRAL WIRE (Blue)
Carries current away from appliance. Stays at a voltage close to zero with respect to earth.

CABLE GRIP
Must be tight to stop the cable moving.

EARTH WIRE (Green + Yellow)

FUSE
Always part of the live wire.

LIVE WIRE (Brown)
Carries current to appliance. Alternates between a positive and negative voltage with respect to neutral terminal.

EARTHING

All appliances with outer metal cases must be earthed. The outer case is connected to the earth pin in the plug through the earth wire ...

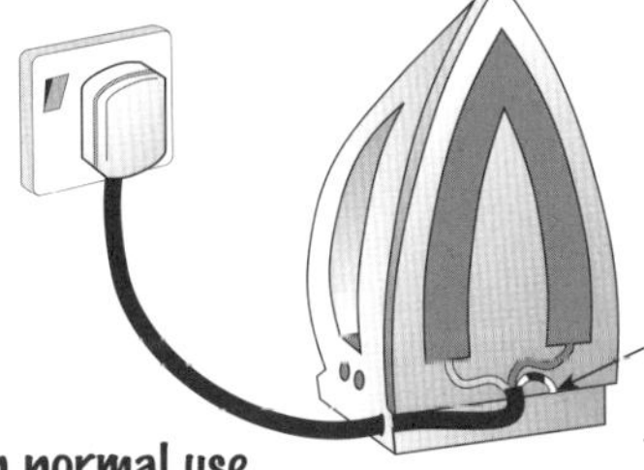

FUSE IN PLUG MELTS

In normal use ...
- ... a current from the live terminal of the mains supply ...
- ... passes through the heating element ...
- ... causing the iron to get hot.

The current then returns through the neutral terminal.

However ...
- ... if a stray wire from the LIVE touches the metal case ...
- ... the case will become live and ...
- ... there is a surge of current to earth ...
- ... via this 'SHORT CIRCUIT' (i.e. the earth wire) which causes ...
- ... the fuse wire to melt, ...
- ... which breaks the circuit.

'LIVE' CASING → SHORT CIRCUIT → CURRENT "SURGES" TO EARTH → FUSE MELTS → CIRCUIT BROKEN

PROTECTION: INSULATION AND DOUBLE INSULATION

ALL ELECTRICAL APPLIANCES should have proper **INSULATION** where ...

... within the plug ... NO FRACTURE OR DAMAGE TO SHEATHING OF WIRES WITH NO BARE WIRE IN VIEW AT THE TERMINALS

... while ...

... between the plug and appliance ... NO FRACTURE OR DAMAGE TO SHEATHING OF CABLE WHERE BARE WIRES MAY SHOW TO MAKE CONTACT WITH USER.

... while ...

... within the appliance ... ALL WIRES AND 'LIVE METAL PARTS' SHOULD BE PROPERLY INSULATED WITH NO CONTACT WITH OUTSIDE CASING.

DOUBLE INSULATION

Some appliances are **DOUBLE INSULATED** where ...
- ... all **METAL PARTS INSIDE** the appliances are **COMPLETELY INSULATED** from ...
- ... any **OUTSIDE PART** of the appliance which **MAY BE HANDLED**.
- These appliances **DO NOT HAVE AN EARTH WIRE** although they are still **PROTECTED BY A FUSE**.

PROTECTION: CIRCUIT BREAKERS

Circuit breakers are **RESETTABLE FUSES** while a ...
... **RESIDUAL CIRCUIT BREAKER** ...

Used with high risk appliances e.g. • ELECTRIC LAWNMOWERS • HEDGE TRIMMERS

- Detects if there is a **DIFFERENCE** between ...
- ... the **CURRENT** in the **LIVE** and **NEUTRAL WIRE** (normally the same).
- When this difference is **GREATER** than a **SAFE LEVEL** ...
- ... the circuit is **BROKEN**.

THIS SWITCHES THE CURRENT OFF MUCH FASTER THAN A FUSE.

KEY POINTS:

- A 3 pin plug consists of a live wire, neutral wire, earth wire, fuse and cable grip.
- All appliances with outer metal cases are earthed.
- Some appliances are double insulated where all metal parts inside the appliance are completely insulated from the outside part of the appliance.

THE ELECTRICITY METER IN YOUR HOME

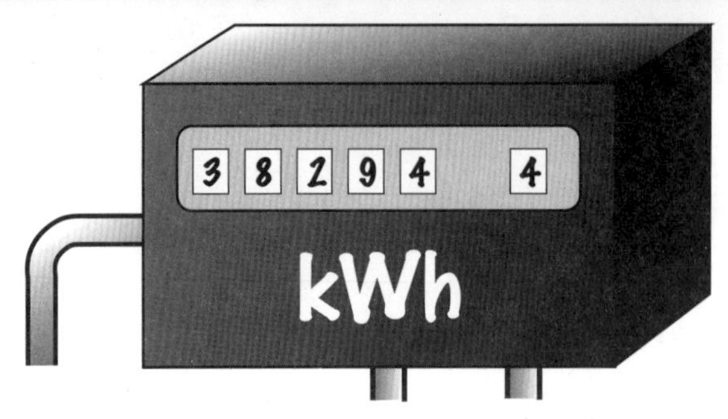

Your meter at home measures the number of Kilowatt-hours of electricity you use. These are sometimes referred to simply as UNITS of electricity, and are used because Joules are too small a unit of energy.

- 1 Kilowatt-hour is equivalent to 3,600,000 Joules!!

THE KILOWATT-HOUR

The Kilowatt-hour is a unit of ENERGY TRANSFERRED and is often called simply a UNIT.

Please remember it is NOT a unit of power- that's the kilowatt!!

A electrical appliance transfers 1kWh of energy if it transfers energy at the rate of 1 kilowatt for one hour.

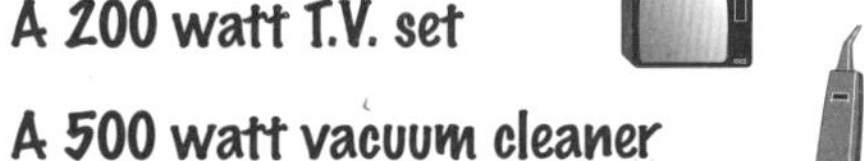

A 200 watt T.V. set ... transfers 1kWh of energy if it is switched on for 5 hours.

A 500 watt vacuum cleaner ... transfers 1kWh of energy if it is switched on for 2 hours.

A 1,000 watt electric fire ... transfers 1kWh of energy if it is switched on for 1 hour.

KILOWATT-HOUR CALCULATIONS

We need to use the following formula ...

UNITS (kWh) = POWER (kW) x TIME (h)

UNITS
P x t

Formula Triangle

EXAMPLE

A 1500 watt electric fire is switched on for 4 hours. How much does it cost if electricity is 9p per unit?

Using the above equation ... UNITS = 1.5(kW) x 4 (hour)
kilowatts remember!
= 6 kilowatt-hours (or UNITS!)

But, **TOTAL COST = NUMBER OF UNITS x COST PER UNIT**

Therefore Total Cost = 6 x 9
= 54 pence

> Remember, to do these calculations, you need ...
> ... to make sure the POWER is in KILOWATTS, AND ...
> ... to make sure that the TIME is in HOURS.

KEY POINTS:

- An electrical appliance transfers 1kWh or 1 Unit of energy if it transfers energy at the rate of 1kW for 1 hour.
- Units (kWh) = Power (kW) x Time (hours).
- Total Cost = Number of Units x Cost per Unit

a.c. and d.c. - comparison as seen on a cathode ray oscilloscope

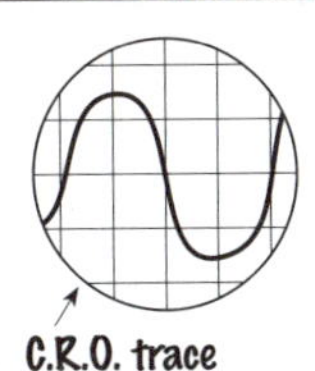

Alternating current (a.c) ....
- Current changes direction of flow ...
- ... back and forth continuously.
- Mains electricity is a.c ...
- ... of frequency 50 Hz ...
- ... i.e. no. of cycles every second.

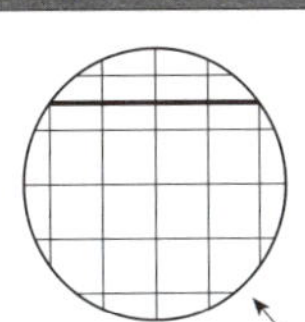

Direct current (d.c) ....
- Current always flows ...
- ... in the same direction.
- Cells and batteries are d.c.

MAKING ELECTRICITY BY ELECTROMAGNETIC INDUCTION (THE DYNAMO PRINCIPLE)

Very simply ...
- If you move a **CONDUCTOR** (WIRE) or a **MAGNET** ...
- ... so that the conductor cuts through the lines of force of the magnetic field ...
- ... then a voltage is induced between the ends of the conductor ...
- ... and a current is induced in the conductor if it is part of a complete circuit.

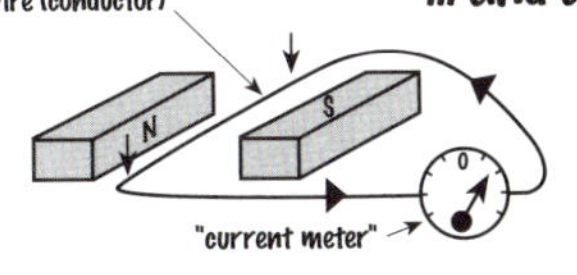

- Moving wire INTO magnetic field ...
- ... induces a current in one direction ...

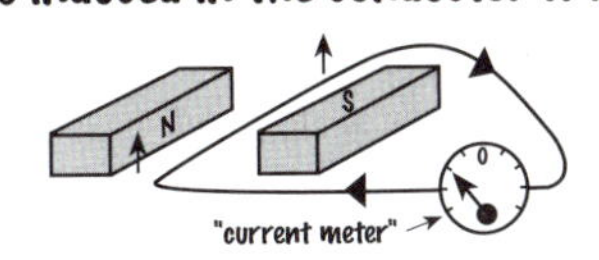

- ... while moving wire OUT OF mag. field ...
- ... induces current in opposite direction ...

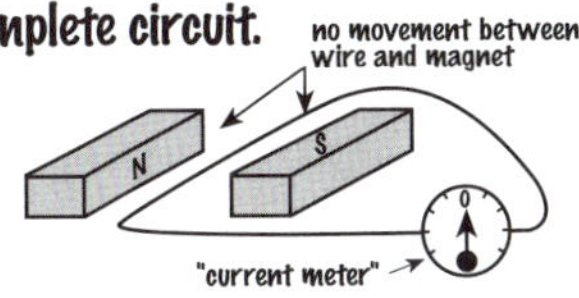

- ... however if there is NO ...
- ... movement of wire or magnet ...
- ... there's no induced current.

THE SAME EFFECT CAN BE SEEN USING A **COIL** AND A **MAGNET** ...

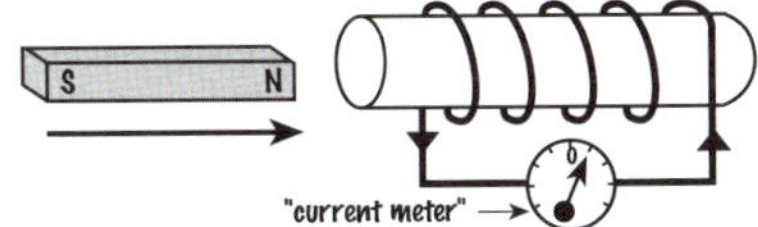

- Moving magnet INTO the coil ...
- ... induces a current in one direction ...

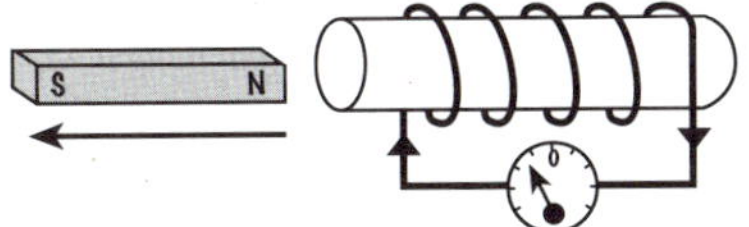

- ... while moving magnet OUT OF the coil ...
- ... induces current in opposite direction ...

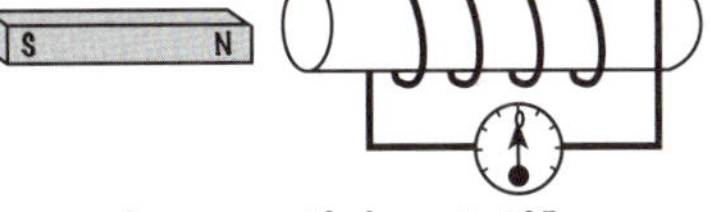

- ... however, if there is NO ...
- ... movement of magnet or coil ...
- ... there's no induced current.

The production of an INDUCED VOLTAGE AND CURRENT using a COIL and MAGNET ...
... is called the DYNAMO PRINCIPLE and forms the basis of mains electricity generation.

THE ALTERNATING CURRENT DYNAMO or GENERATOR

... uses the dynamo principle to INDUCE current ...
... which **reverses direction each revolution,** ...
... as the coil is driven across the lines of magnetic field.

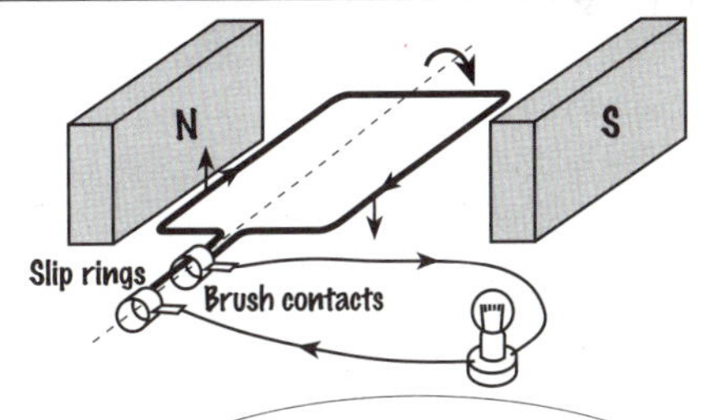

So the general idea for generating electricity is to get the wire crossing the lines of force of the magnetic field as often as possible in order to induce more current.

N.B. The BRUSH CONTACTS are spring-loaded so that they push gently against the SLIP RINGS so that the circuit remains complete. Gradually they wear away and have to be replaced.

This is done by:-

- INCREASING THE SPEED OF MOVEMENT
- INCREASING THE MAGNETIC FIELD STRENGTH
- INCREASING THE NUMBER OF TURNS ON THE COIL
- INCREASING THE AREA OF THE COIL

INCREASING THESE FOUR THINGS INCREASES THE NUMBER OF LINES OF MAGNETIC FIELD CUT PER SECOND, AND THEREFORE INCREASES THE CURRENT.

(Remember this by "Sloppy Alternator Fails Test")

You will not be asked to draw the a.c dynamo but you must be able to explain its construction and performance.

KEY POINTS:

- A voltage is induced between the ends of a conductor and a current is induced in the conductor if it is part of a complete circuit when it cuts through a magnetic field.

DISTANCE TRAVELLED IN METRES, ... EVERY SECOND

The relationship: AVERAGE SPEED (m/s) = DISTANCE (m) / TIME TAKEN (s)

$$\text{AVERAGE SPEED (m/s)} = \frac{\text{DISTANCE (m)}}{\text{TIME TAKEN (s)}}$$

FORMULA TRIANGLE: d / s x t

Usually measured in m/s, Km/h or miles/h

N.B. **VELOCITY OF AN OBJECT IS ITS SPEED IN A PARTICULAR DIRECTION**

YOU MUST KNOW THIS FORMULA - SO LEARN IT

CONSTANT or UNIFORM SPEED

- e.g. cyclist travelling at 8m/s

- Same DISTANCE ... every SECOND ... CONSTANT SPEED.

EXAMPLE: "Calculate the average speed of a cyclist who travels 2400m in 5mins."

AVERAGE SPEED = DISTANCE TRAVELLED / TIME TAKEN = 2400(m) / 300 (secs remember!) = 8m/s

DISTANCE - TIME GRAPHS

1. STATIONARY

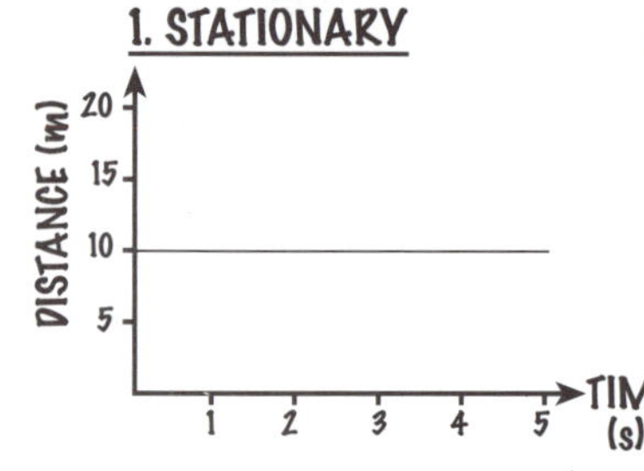

2. MOVING AT CONSTANT SPEED

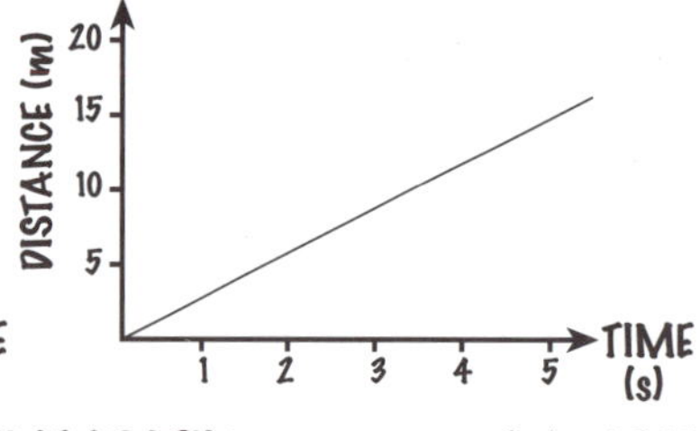

3. MOVING AT GREATER CONSTANT SPEED

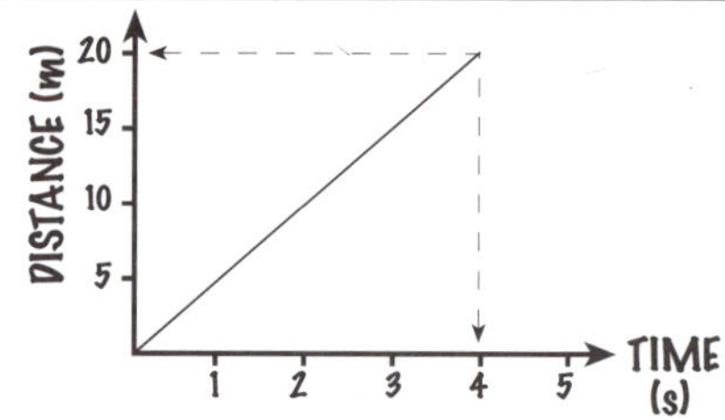

- The SLOPE of a DISTANCE-TIME GRAPH is a measure of the SPEED OF THE OBJECT ...
- ... and the STEEPER the SLOPE, the GREATER the SPEED.

EXAMPLE

For No. 3, GRADIENT = SPEED = DISTANCE TRAVELLED / TIME TAKEN = 20m / 4s = 5m/s (see graph above).

DISTANCE - TIME GRAPHS FOR CHANGING VELOCITY AND ACCELERATION

1. CHANGING VELOCITY

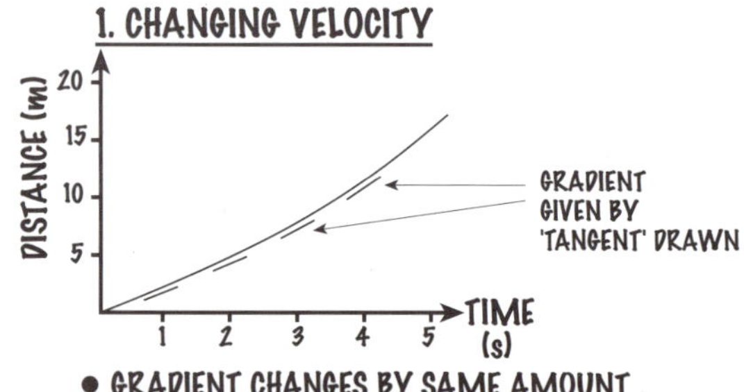

- GRADIENT CHANGES BY SAME AMOUNT ...
- ... EVERY SECOND and so
- VELOCITY CHANGES BY SAME AMOUNT ...
- ... EVERY SECOND
- OBJECT MOVING WITH CONSTANT CHANGING VELOCITY.

2. CHANGING ACCELERATION

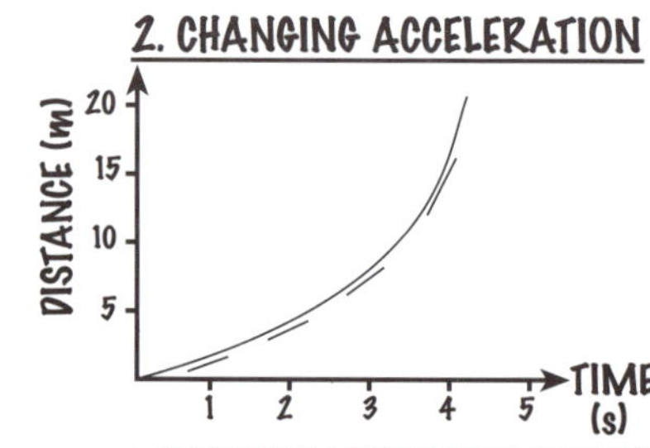

- GRADIENT CHANGES BY DIFFERENT AMOUNTS ...
- ... EVERY SECOND and so
- VELOCITY CHANGES BY DIFFERENT AMOUNTS ...
- ... EVERY SECOND
- OBJECT MOVING WITH CHANGING ACCELERATION.

KEY POINTS:

- Speed = Distance ÷ Time taken
- Measured in m/s, Km/h or miles/h.
- Velocity is speed in a particular direction.
- Distance-time graphs can be used to represent the motion of an object.

MASS, WEIGHT AND GRAVITY

Basically ...
- The MASS in KILOGRAMS (Kg) of an OBJECT is the AMOUNT OF MATTER it contains ...
- ... and the WEIGHT in NEWTONS (N) of an OBJECT is the DOWNWARD FORCE ...
- ... on that object due to GRAVITY.

Where ...
- GRAVITY is the FORCE of ATTRACTION that exists between ANY TWO OBJECTS ...
- ... in this case the OBJECT and the EARTH.

While ...
- Near the surface of the earth EVERY 1KG OF MATTER has a WEIGHT of 10N.

ACCELERATION DUE TO GRAVITY

The acc^n of a FREELY FALLING OBJECT on earth depends on whether the object is falling through ...

- A VACUUM.
- AIR or ANOTHER FLUID (e.g. water).

SINCE ALL FALLING OBJECTS EXPERIENCE A DOWNWARDS FORCE (GRAVITY) ...
- ... THEN WHILE FALLING THEY MUST ALSO BE ACCELERATING however ...

- OBJECT DOES NOT experience ...
- ... an OPPOSING UPWARD FORCE ...
- ... i.e. FRICTION and so ...
- ... ALL objects fall with a ...
- ... CONSTANT ACC^n in a VACUUM.
- ... Its value is $10m/s^2$

... in AIR and OTHER FLUIDS

- OBJECT DOES experience ...
- ... an OPPOSING UPWARD FORCE ...
- ... due to FRICTION between the OBJECT ...
- .... and the AIR or FLUID it falls through.
- Therefore it will ONLY ACCELERATE UP TO A POINT! (see below).

TERMINAL VELOCITY - a freefalling experience

This is the CONSTANT VELOCITY reached by FALLING OBJECTS in AIR and OTHER FLUIDS (not in a VACUUM!) and so ...

The next time you jump out of a plane remember ... the FORCES acting on you are ...
- WEIGHT, W (↓) and
- FRICTION DUE TO AIR called AIR RESISTANCE or DRAG, R (↑).

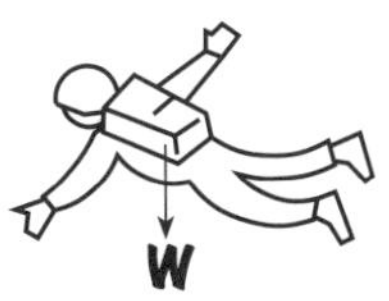

- INITIALLY ...
- ... only force acting is W ...
- ... air resistance R, experienced ...
- ... as soon as you fall!

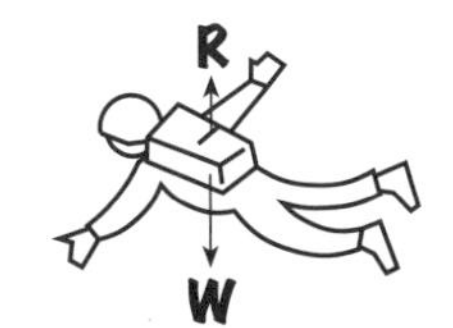

- As you fall, you accelerate ...
- ... because $W > R$
- However, as you accelerate, ...
- ... R increases, but W doesn't!

R

W

- EVENTUALLY ...
- R increases until $R = W$ ...
- ... meaning no resultant force so, ...
- ... no acceleration. TERMINAL VELOCITY REACHED

Factors which affect terminal velocity

1. AIR RESISTANCE
- the GREATER the AIR RESISTANCE ...
- ... the MORE RESISTANCE there is AGAINST MOVEMENT.
- It then takes LESS TIME until R=W which means ...
- ... object ACCELERATING FOR A SHORTER TIME and so ...
- ... TERMINAL VELOCITY of object is REDUCED.

2. SHAPE OF OBJECT
- STREAMLINED OBJECTS offer ...
- ... LESS RESISTANCE AGAINST MOVEMENT.
- It then takes MORE TIME until R=W which means ...
- ... object ACCELERATING FOR A LONGER TIME and so ...
- ... TERMINAL VELOCITY of object is INCREASED.

KEY POINTS:

- Gravity gives an object weight.
- A mass of 1kg has a weight of 10N near the surface of the earth.
- An object falling through a medium (e.g. air) will reach terminal velocity.

TURNING EFFECT OF A FORCE - MOMENTS

Depends on
- SIZE OF APPLIED FORCE - greater force, greater turning effect.
- PERPENDICULAR DISTANCE FROM PIVOT - greater distance, greater turning effect.

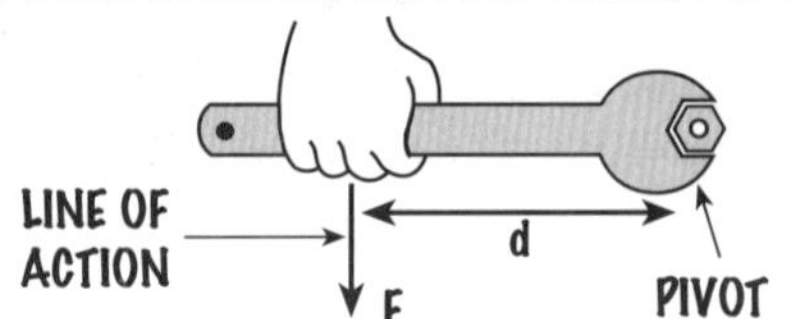

The relationship ...

MOMENT (Nm) = FORCE (N) x PERPENDICULAR DISTANCE FROM PIVOT (m)

Remember! The standard unit is the <u>NEWTON - METRE</u>.

<u>Principle of Moments</u> - balanced moments

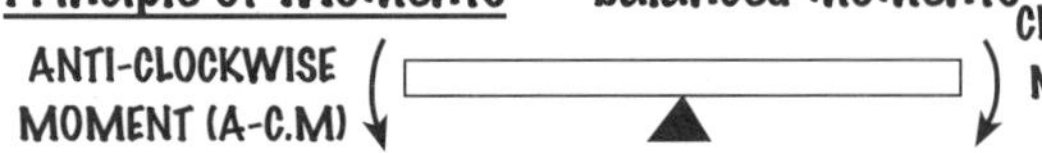

... and when balanced, **TOTAL CLOCKWISE MOMENT = TOTAL ANTICLOCKWISE MOMENT**

<u>Examples:</u>

1) Calculate the unknown force if the see-saw below is balanced

A-C.M 2m 4m 10N F C.M

When balanced, TOTAL C.M = TOTAL A-C.M

$F \times 4m = 10N \times 2m$

$4F = 20$

$F = \frac{20}{4} = \underline{5N}$

BOTH THESE FORMULAE ARE GIVEN

2) Two girls weighing 250N and 300N sit on one side of a seesaw 1m and 1.5m from the pivot respectively. Where must a boy weighing 350N sit in order to balance the seesaw?

A diagram is essential.

A-C.M d 1.5m 1m 350N 250N 300N C.M

When balanced, TOTAL C.M = TOTAL A-C.M

$(250N \times 1m) + (300N \times 1.5m) = 350N \times d$

$250 + 450 = 350d$

$700 = 350d \quad d = \frac{700}{350} = \underline{2m}$

EFFECT OF APPLIED FORCES ON SOLID OBJECTS - HOOKE'S LAW

APPLIED FORCES acting on SOLID OBJECTS can result in ...
- EXTENSION ... where the object is STRETCHED.
- COMPRESSION ... where the object is SQUASHED.

Here we will consider in more detail the EXTENSION PRODUCED due to STRETCHING as the result of an APPLIED FORCE on an OBJECT e.g. METAL WIRE or SPRING. A typical graph of EXTENSION against APPLIED FORCE would look like:

EXTENSION

APPLIED FORCE or LOAD

OBEYS HOOKE'S LAW

NO LONGER OBEYS HOOKE'S LAW

OBJECT SUFFERS DAMAGE (LARGER INCREASE IN EXTENSION THAN EXPECTED).

ELASTIC LIMIT (OBJECT RETAINS ORIGINAL SIZE UP TO THIS POINT).

OBJECT IN THIS REGION RETAINS ORIGINAL SIZE, WHEN APPLIED FORCE IS REMOVED.

<u>Hooke's Law</u>

Up to the ELASTIC LIMIT ...
- ... EXTENSION IS DIRECTLY PROPORTIONAL TO APPLIED FORCE ...
- ... i.e. extension is doubled if we double the applied force etc.
- This relationship between EXTENSION and
- APPLIED FORCE ...
- ... is known as <u>HOOKE'S LAW.</u>

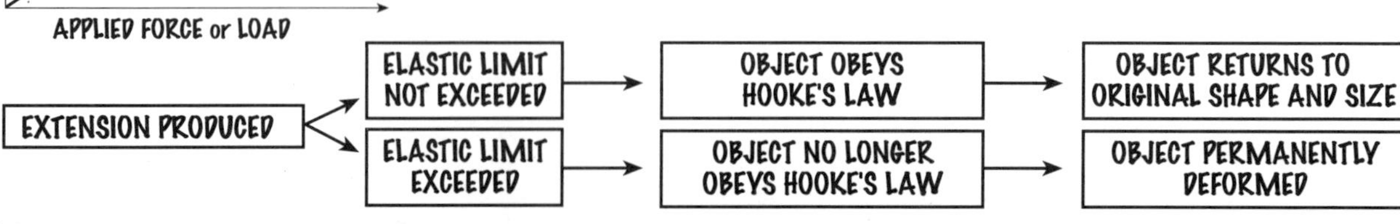

KEY POINTS:

- Moment (Nm) = Force (N) x Perpendicular distance from pivot (m).
- For balanced moments, Total clockwise moment = Total anticlockwise moment
- If elastic limit is not exceeded a stretched object obeys Hooke's Law.

WAVES AND THE ELECTROMAGNETIC SPECTRUM

WAVES are
- ... a REGULAR PATTERN OF DISTURBANCE ...
- ... which TRANSFER ENERGY from one point to another ...
- ... without any TRANSFER OF MATTER.

FEATURES OF WAVES - some important definitions

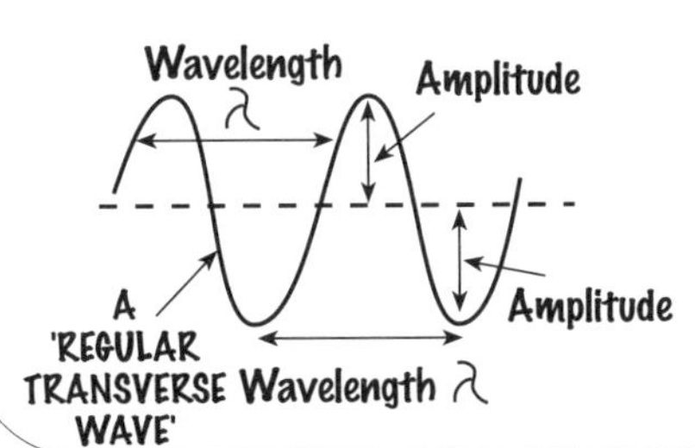

AMPLITUDE (UNITS: cm or m)
- ... is the MAXIMUM DISTURBANCE, ...
- ... i.e. the GREATEST DISPLACEMENT ...
- ... from the MEAN POSITION

WAVELENGTH (UNITS: m).
- ... is the DISTANCE between ...
- ... CONSECUTIVE CRESTS ...
- ... or TROUGHS (symbol = λ)

FREQUENCY (UNITS: Hertz)
- ... is the NUMBER OF WAVES PRODUCED ...
- ... or PASSING A PARTICULAR POINT ...
- ... in ONE SECOND.

THE ELECTROMAGNETIC SPECTRUM

Some facts-
- ALL ELECTROMAGNETIC WAVES OF DIFFERENT WAVELENGTH AND FREQUENCY ...
- TRAVEL AT THE SAME SPEED (300,000,000 m/s) THROUGH THE AIR ...
- ... OR THROUGH A VACUUM.

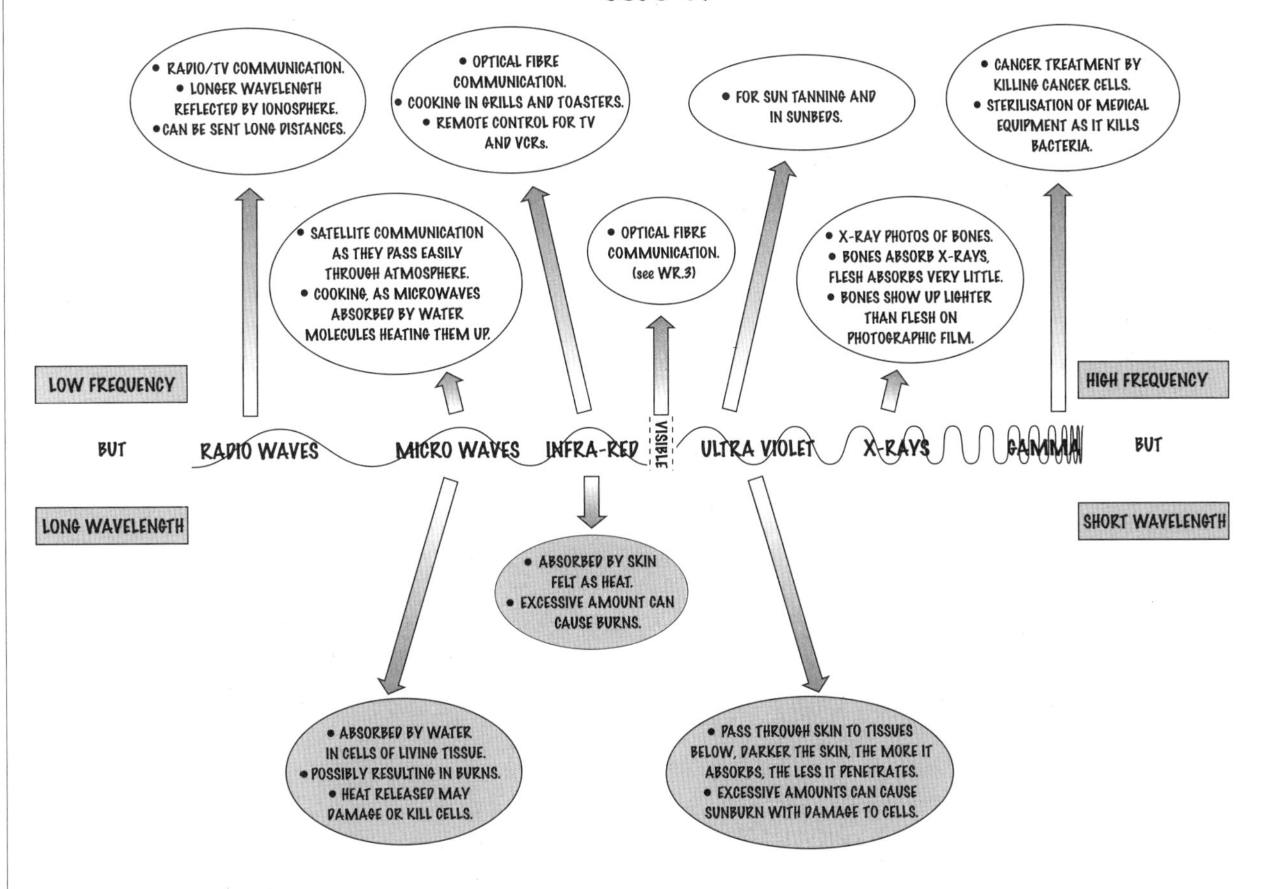

KEY POINTS:

- Two types of wave: Transverse and Longitudinal.
- Amplitude, Wavelength and Frequency are all features of waves.
- Radiowaves, Microwaves, Infra-red, Visible, Ultra Violet, X-rays and Gamma rays make up the E-M spectrum.

LIGHT

VISIBLE LIGHT is a small part of the ELECTROMAGNETIC SPECTRUM (see FT.11).

Light:
- Travels in STRAIGHT LINES e.g. formation of SHADOWS.
- Travels at a CONSTANT SPEED through a UNIFORM MEDIUM ...
  ... the DENSER the MEDIUM, the SLOWER its SPEED.
- Can also travel through a VACUUM.

REFLECTION OF LIGHT

This occurs when light strikes a SURFACE or a MIRROR resulting in it CHANGING ITS DIRECTION.

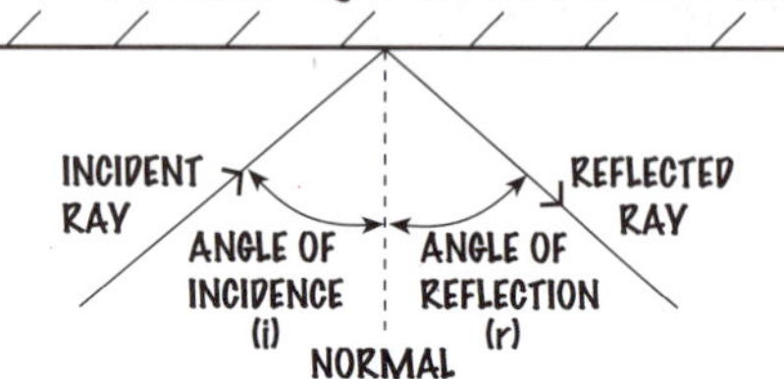

Always -

ANGLE OF INCIDENCE (i) = ANGLE OF REFLECTION (r)

REFRACTION OF LIGHT

- Light CHANGES DIRECTION, when it CROSSES A BOUNDARY ...
- ... between ONE TRANSPARENT MEDIUM AND ANOTHER (OF DIFFERENT DENSITY) ...
- .. unless it meets the boundary along a NORMAL (AT 90°)

GLASS OR PERSPEX
AIR

- Rays of light are REFRACTED ..
- ... TOWARDS the NORMAL because ...
- ... they pass from a LESS DENSE ...
- ... to a MORE DENSE MEDIUM.

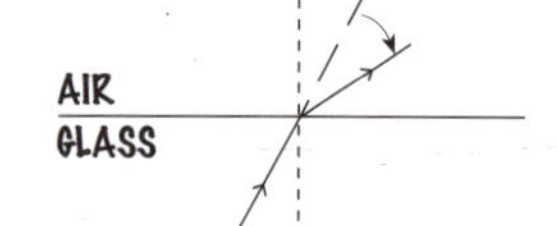

- Rays of light are REFRACTED ...
- ... AWAY from the NORMAL because ...
- ... they pass from a MORE DENSE ...
- ... to a LESS DENSE MEDIUM.

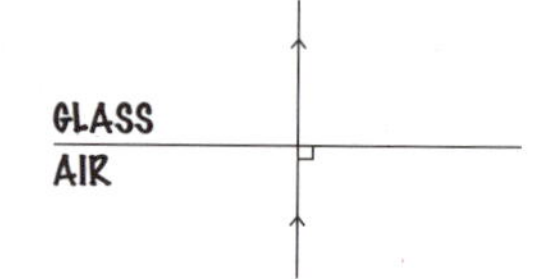

- Rays of light NOT DEVIATED since ...
- ... they meet boundary ...
- ... ALONG THE NORMAL (at 90°)

REFRACTION OCCURS BECAUSE LIGHT CHANGES SPEED WHEN IT PASSES FROM ONE MEDIUM INTO ANOTHER where ...
- LESS DENSE to MORE DENSE ... it SLOWS DOWN.
- MORE DENSE to LESS DENSE ... it SPEEDS UP.

DIFFRACTION OF LIGHT

- When WAVES MOVE THROUGH A GAP or PASS AN OBSTACLE ...
- ... they SPREAD OUT FROM THEIR EDGES.

Diffraction is most obvious when:

1.) SIZE OF GAP IS SIMILAR TO WAVELENGTH OF WAVES.

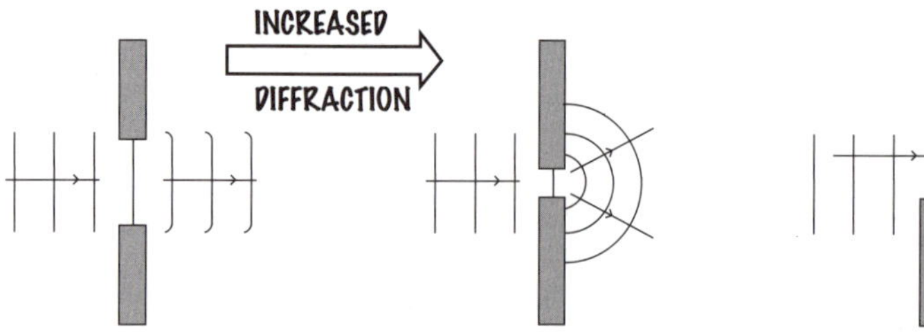

2.) WAVES WHICH PASS OBSTACLES HAVE LONG WAVELENGTHS.

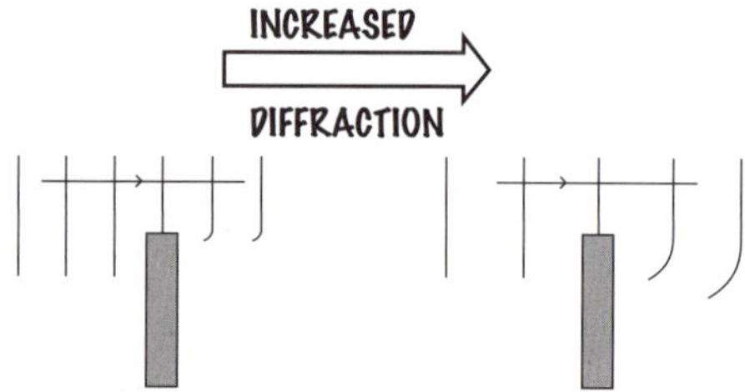

The size of gap has to be very, very small because light has a very small wavelength.

KEY POINTS:

- Light is a transverse wave.
- Light can be reflected, refracted and diffracted.

FORCE AND TRANSFERS SUMMARY QUESTIONS

1. What is current?
2. Explain how a) an AMMETER and b) a VOLTMETER must be connected in a circuit.
3. Name three appliances which use the heating effect in resistors.
4. Give one fact about the CURRENT and VOLTAGE in a SERIES CIRCUIT.
5. Give one fact about the CURRENT and VOLTAGE in a PARALLEL CIRCUIT.
6. Give one advantage and one disadvantage of having light bulbs in a SERIES CIRCUIT.
7. Give one advantage and one disadvantage of having light bulbs in a PARALLEL CIRCUIT.
8. What is resistance?
9. If the current through a component is 4 amps at a voltage of 8 volts, what is its RESISTANCE? (remember the units)
10. Draw a current - voltage graph for a) a RESISTOR at constant temperature b) a FILAMENT BULB and c) a DIODE.
11. What is power?
12. What units do we use for POWER?
13. What is a fuse?
14. Work out the fuse needed for an appliance with a power rating of 1kW. Assume 240 volts.
15. Name the three wires in a three pin plug and say what colour they are.
16. What is DOUBLE INSULATION?
17. How does a RESIDUAL CIRCUIT BREAKER work?
18. What is a) a KILOWATT and b) a KILOWATT-HOUR?
19. Is a Kilowatt-hour a unit of power or energy?
20. A 1000W electric fire is used for 5 hours. What is the cost at 7p per unit?
21. A 2000W electric shower is used for 30 minutes. What is the cost at 7p per unit?
22. What is the difference between A.C. and D.C.
23. Describe how a voltage can be INDUCED across a wire.
24. Describe how a generator may be made to induce more current.
25. A boy runs 1200m in 5 minutes. What is his average speed?
26. What does the gradient of a distance-time graph represent?
27. What is a) mass b) weight and c) gravity?
28. What are the names of the forces which act on a falling object?
29. If an object reaches terminal velocity, which two forces are balanced?
30. What is a moment?
31. What are the principle of moments?
32. Draw a labelled diagram of extension against applied force for an elastic object.
33. What is the elastic limit?
34. Draw a graph of extension against applied force for a material using the following results.

Applied Force (N)	0	10	20	30	40	50	60	70
Extension (mm)	0	5	10	15	20	25	31	39

On your graph mark in a) region where material retains original size b) elastic limit c) region material suffers permanent damage.

35. What is a) amplitude b) wavelength c) frequency?
36. What is a) amplitude b) wavelength c) frequency measured in?
37. Draw a diagram to show the reflection of light.
38. What two things are always the SAME when light is reflected?
39. Draw a diagram to show the refraction of light.
40. Why does refraction occur?

ENERGY SOURCES

ENERGY RESOURCES I - Non-renewables

E 1

- NON-RENEWABLE ENERGY RESOURCES are those that will ONE DAY RUN OUT and ...
- ... once they have been used they CANNOT BE USED AGAIN.

THE OPTIONS:-

NON - RENEWABLES

FOSSIL FUELS

COAL | OIL | NATURAL GAS | NUCLEAR FUEL

They are ALL used in the GENERATION of ELECTRICITY where ...
... THE **FUEL** IS USED TO GENERATE **HEAT**, WHICH THEN BOILS WATER TO MAKE **STEAM** TO DRIVE THE **TURBINES**, WHICH TURN THE **GENERATORS** PRODUCING **ELECTRICITY**.

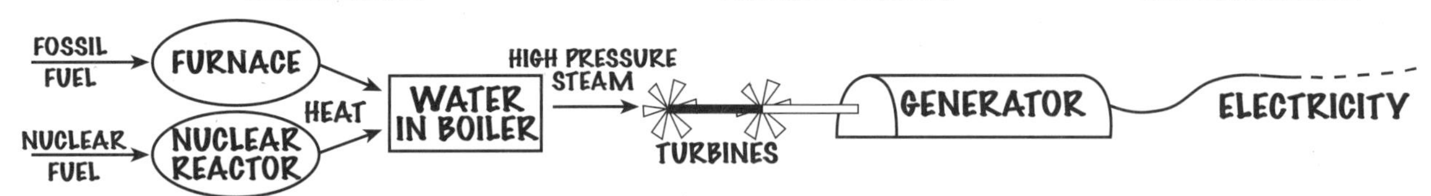

CONSERVATION OF NON - RENEWABLE ENERGY RESOURCES

Over 90% of the electricity produced in the world is produced by using non-renewable resources.
There is a growing awareness that CONSERVATION of these resources is needed for many reasons:-

	COAL (FOSSIL FUEL)	OIL/GAS (FOSSIL FUELS)	NUCLEAR (URANIUM + PLUTONIUM)
FINITE NATURE	ONLY 100 YRS. WORTH OF COAL LEFT.	PERHAPS 30 YRS. OF OIL/GAS LEFT.	GOOD SUPPLIES BUT IS DIFFICULT TO OBTAIN.
ENVIRONMENTAL PROBLEMS	GLOBAL WARMING, ACID RAIN (CAN BE REDUCED BY REMOVING SULPHUR DIOXIDE AFTER BURNING, CALLED 'SCRUBBING')	GLOBAL WARMING TANKER SPILLAGE	RADIOACTIVE SUBSTANCES MAY ESCAPE (SITE THE POWER STATION IN 'REMOTE' LOCATIONS WITH MANY FAIL SAFE DEVICES)
ETHICAL CONSIDERATIONS	GLOBAL WARMING IS CAUSING AN INCREASE IN THE TEMP. OF THE EARTH. ACID RAIN CAUSES DAMAGE TO TREES, PLANTS, FISH AND BUILDINGS.	GLOBAL WARMING IS CAUSING AN INCREASE IN THE TEMP. OF THE EARTH. ANY TANKER SPILLAGE HAS A DEVASTATING EFFECT ON MARINE LIFE.	DISCHARGE OF RADIOACTIVE LIQUIDS INTO THE SEA WILL AFFECT THE FOOD CHAIN. WASTE PRODUCTS NEED VERY SAFE STORAGE FOR THOUSANDS OF YEARS.
CONCERNS FOR THE FUTURE	THERE IS A CONCERN THAT ALL PEOPLE, BOTH NOW AND IN THE FUTURE SHOULD HAVE A FAIR AND APPROPRIATE SHARE OF THE EARTH'S RESOURCES. APART FROM GENERATING ELECTRICITY, FOSSIL FUELS ESPECIALLY, HAVE MANY OTHER VALUABLE USES (MAKING OF PLASTICS etc) AND FOR TRANSPORT (OIL).		

KEY POINTS:

- Coal, Oil, Natural Gas (the fossil fuels) and Nuclear fuel are non-renewable energy resources.

ENERGY RESOURCES II - Renewables

- RENEWABLE ENERGY RESOURCES are those that will NOT RUN OUT and ...
- ... are CONTINUALLY BEING REPLACED.

THE OPTIONS:-

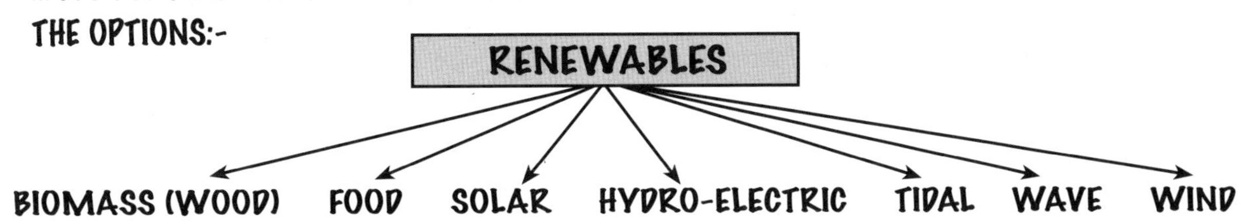

THESE CAN BE USED DIRECTLY ...

- PLANTS, ESPECIALLY TREES, CAN BE GROWN TO PROVIDE FUEL FOR HEATING.
- CROPS CAN BE GROWN AND HARVESTED TO PROVIDE FOOD SUPPLIES.

These are ALL used in the GENERATION of ELECTRICITY where ...
... THE **ENERGY RESOURCE** IS USED TO DRIVE **TURBINES DIRECTLY** etc ...
... WITH THE EXCEPTION OF SOLAR ...
... SO NO NASTY BURNING IS INVOLVED.

WIND or WATER → TURBINES → GENERATOR → ELECTRICITY

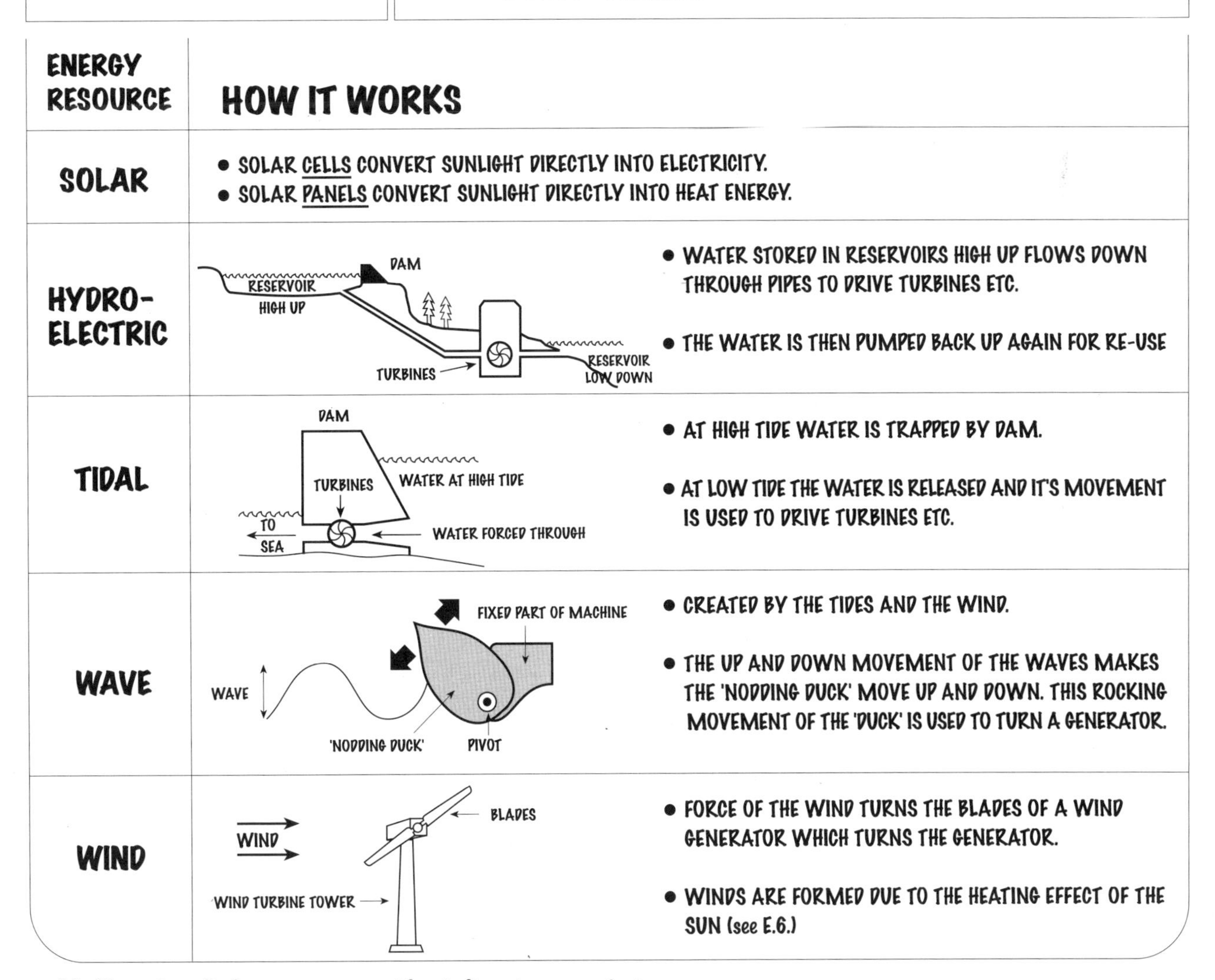

ENERGY RESOURCE	HOW IT WORKS
SOLAR	• SOLAR CELLS CONVERT SUNLIGHT DIRECTLY INTO ELECTRICITY. • SOLAR PANELS CONVERT SUNLIGHT DIRECTLY INTO HEAT ENERGY.
HYDRO-ELECTRIC	• WATER STORED IN RESERVOIRS HIGH UP FLOWS DOWN THROUGH PIPES TO DRIVE TURBINES ETC. • THE WATER IS THEN PUMPED BACK UP AGAIN FOR RE-USE
TIDAL	• AT HIGH TIDE WATER IS TRAPPED BY DAM. • AT LOW TIDE THE WATER IS RELEASED AND IT'S MOVEMENT IS USED TO DRIVE TURBINES ETC.
WAVE	• CREATED BY THE TIDES AND THE WIND. • THE UP AND DOWN MOVEMENT OF THE WAVES MAKES THE 'NODDING DUCK' MOVE UP AND DOWN. THIS ROCKING MOVEMENT OF THE 'DUCK' IS USED TO TURN A GENERATOR.
WIND	• FORCE OF THE WIND TURNS THE BLADES OF A WIND GENERATOR WHICH TURNS THE GENERATOR. • WINDS ARE FORMED DUE TO THE HEATING EFFECT OF THE SUN (see E.6.)

You'll notice all the non-renewables belt out energy but ...
... are environmentally disastrous as they are slowly poisioning the earth while ...
... the renewables are very 'earth friendly' but can't meet the demand. IS THERE A WINNER?

KEY POINTS:

- Biomass, Food, Solar, Hydro-electric, Tidal, Wave and Wind are renewable energy resources.

THE SUN'S ENERGY INPUT TO THE EARTH

The SUN is the ORIGINAL SOURCE of MOST of the EARTH'S ENERGY RESOURCES ...
... both NON - RENEWABLE and RENEWABLE

e.g. formation of coal
REMEMBER!! THIS PROCESS TOOK PLACE MILLIONS OF YEARS AGO.

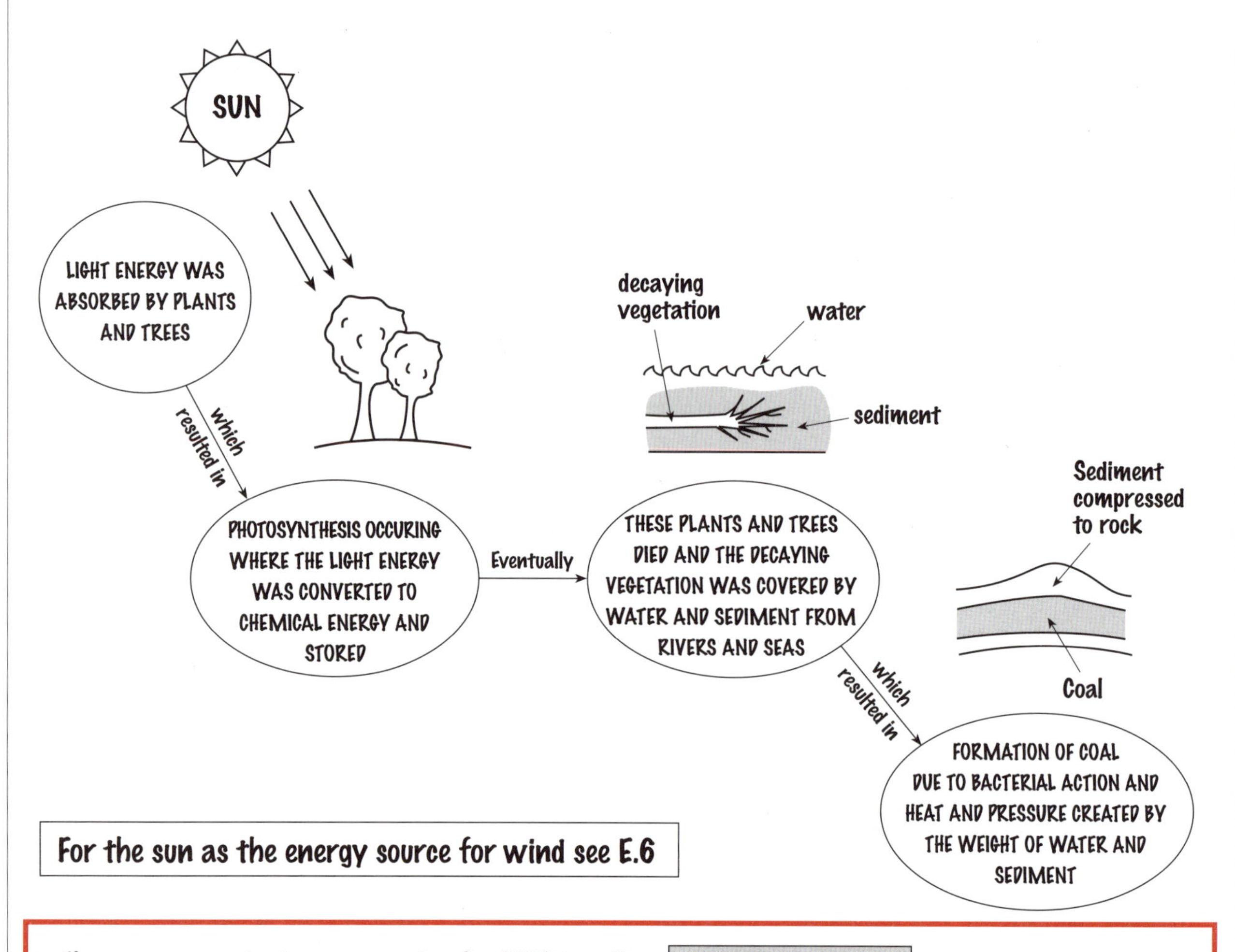

For the sun as the energy source for wind see E.6

The energy producing process in the SUN is called NUCLEAR FUSION ...

At centre of SUN, HYDROGEN nuclei at GREAT PRESSURE and VERY HIGH TEMP ... → ... Join together ... → ... to form HELIUM NUCLEI which have a SMALLER MASS than the HYDROGEN NUCLEI that FUSED TOGETHER to form them ... → ... so ... → ... this 'LOST MASS' is CONVERTED into HEAT ENERGY. Billions of these reactions happen every second.

KEY POINTS:

- The Sun is the original source of most non-renewable and renewable energy resources.
- The energy producing process in the Sun is called Nuclear Fusion.

ENERGY TRANSFER IN REACTIONS

Whenever chemical reactions occur, energy is usually transferred to or from the surroundings, usually as heat.

EXOTHERMIC AND ENDOTHERMIC REACTIONS

Very simply ...

Chemical reaction where HEAT ...
... is TRANSFERRED TO SURROUNDINGS ...
... usually indicated ...
... by a **RISE** in TEMPERATURE is ... **EXOTHERMIC**

... while ...

Chemical reaction where HEAT ...
... is TRANSFERRED FROM THE SURROUNDINGS ...
... usually indicated ...
... by a **FALL** in TEMPERATURE is ... **ENDOTHERMIC**

However, it's not quite as simple as this and here are ...
... TWO massively important facts about chemical reactions ...

- ... Energy must be SUPPLIED TO BREAK BONDS IN REACTANT PARTICLES ...
- ... Energy is RELEASED WHEN BONDS ARE FORMED IN PRODUCT PARTICLES.

<u>So, in EXOTHERMIC REACTIONS</u> ...

ENERGY NEEDED TO BREAK EXISTING BONDS	<	ENERGY RELEASED FROM FORMING NEW BONDS	So, HEAT IS TRANSFERRED TO SURROUNDINGS

<u>But, in ENDOTHERMIC REACTIONS</u> ...

ENERGY NEEDED TO BREAK EXISTING BONDS	>	ENERGY RELEASED FROM FORMING NEW BONDS	So, HEAT IS TAKEN IN FROM SURROUNDINGS

ENERGY LEVEL DIAGRAMS - Reaction of methane with oxygen

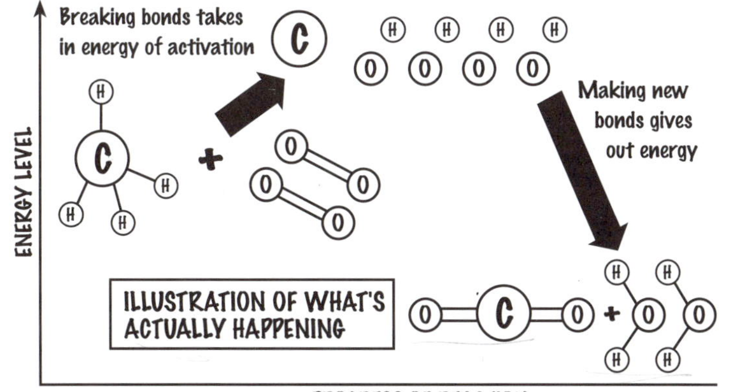

ENERGY LEVEL DIAGRAM

ENERGY LEVEL

$CH_4 + 2O_2$ reactants

ACTIVATION ENERGY

HEAT OF REACTION (-ve)

$CO_2 + 2H_2O$ products

PROGRESS OF REACTION

The diagrams above illustrate ...

- ... the NEED FOR ACTIVATION ENERGY - the minimum energy to break bonds and start the reaction off.
- ... the HEAT OF REACTION - the difference between energy of the products and energy of reactants.

HEAT OF REACTION IS NEGATIVE FOR EXOTHERMIC REACTIONS AND POSITIVE FOR ENDOTHERMIC ONES.

<u>TWO MORE EXAMPLES OF ENERGY LEVEL DIAGRAMS:-</u>

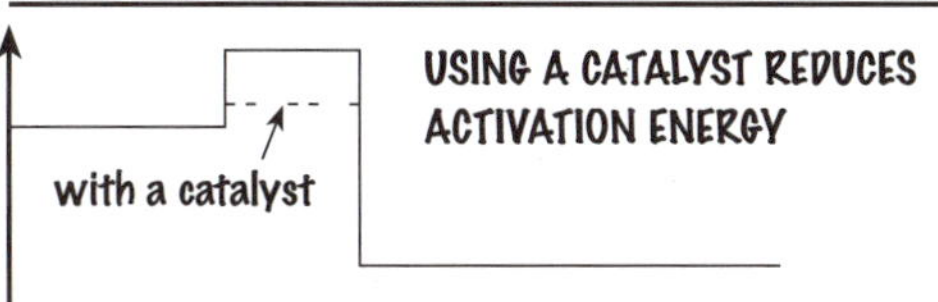

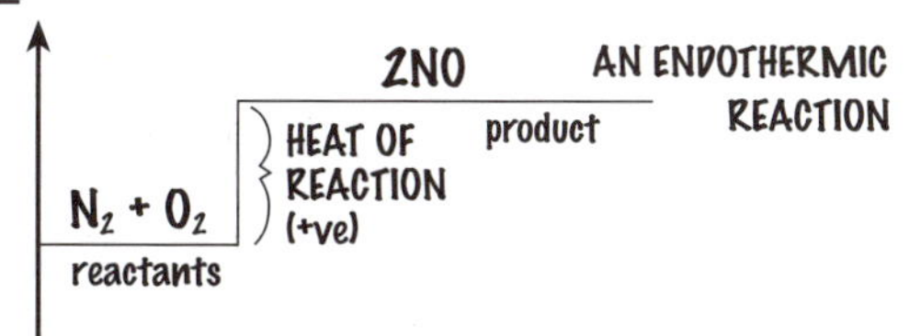

KEY POINTS:

- A chemical reaction where heat is transferred to the surroundings is exothermic.
- A chemical reaction where heat is transferred from the surroundings is endothermic.
- The activation energy is the minimum energy needed to break bonds and to start the reaction off.
- The heat of reaction is the difference between the energy of the products and the energy of the reactants.

TYPES OF ENERGY TRANSFER

ENERGY exists in many FORMS. ENERGY TRANSFERS involve the transfer of energy TO and FROM the following forms ...

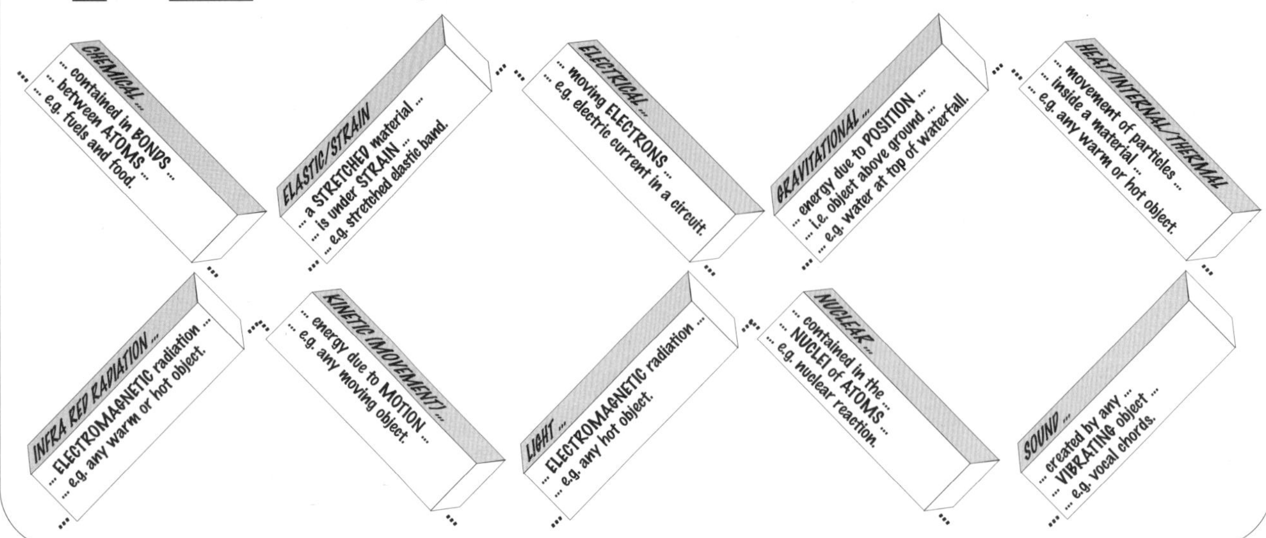

EFFICIENCY OF ENERGY TRANSFER

Whenever energy is transferred from one form to another some is 'wasted' - usually as heat and often sound, scientists are constantly working to reduce this wastage.
They want to make energy transfer devices more **EFFICIENT** .
Here are FOUR examples of the intended energy transfer and wastage in everday devices:

1. Tungsten filament light bulb

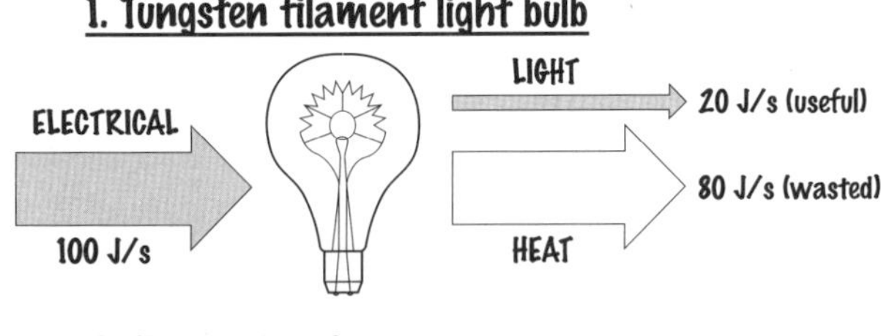

2. Low energy light bulb

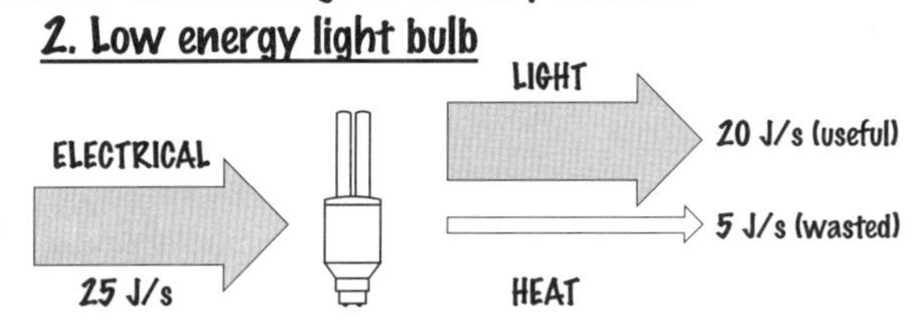

3. Electric kettle

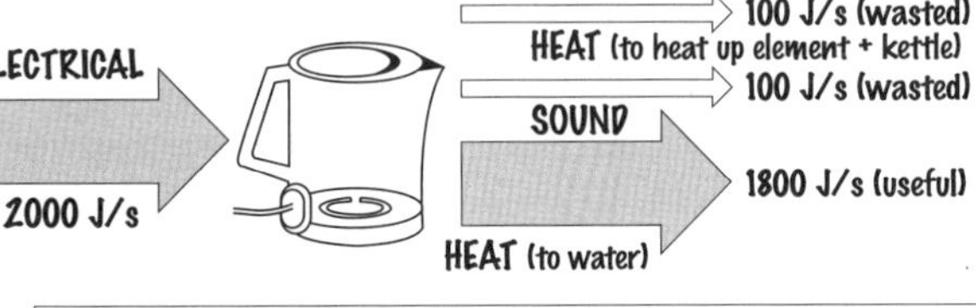

4. Electric motor (shown in a hairdrier)

The EFFICIENCY of any transfer device is given by this relationship:

$$\text{EFFICIENCY} = \frac{\text{USEFUL ENERGY OUTPUT}}{\text{TOTAL ENERGY INPUT}}$$

So, for the filament light bulb above, Efficiency $= \frac{20}{100} \times 100 = 20\%$

and the low energy light bulb, Efficiency $= \frac{20}{25} \times 100 = 80\%$

(× 100: to convert it to a percentage)

... WHICH ONE WOULD YOU BUY?

You will be given this formula in the exam but you must know how to use it!!

KEY POINTS:

- Chemical, Infra red, Elastic, Kinetic, Electrical, Light, Gravitational, Nuclear, Heat and Sound are all forms of energy.
- Efficiency = Useful Energy Output ÷ Total Energy Input.

OTHER ENERGY TRANSFERS II - Conduction and Convection

Heat energy is transferred as a result of temperature differences. You need to understand the following four methods of heat energy transfer ...

1. CONDUCTION

- This is the transfer of heat energy through a substance **WITHOUT THE SUBSTANCE MOVING.**
- **METALS** are very good conductors ... **NON-METALS** are usually poor conductors (INSULATORS) ... and **GASES** are very poor conductors.
- As the metal is heated **FREE ELECTRONS** move more quickly ...
- ... causing them to **DIFFUSE** through the metal ...
- ... colliding with other electrons and passing on their energy.
- These in turn collide with other electrons.

2. CONVECTION

- This is the transfer of heat energy through ...
- ... a **LIQUID** or a **GAS** because they can **FLOW.**
- The particles in a liquid or gas MOVE MORE ENERGETICALLY ... ... when **HOT** which causes it to increase in volume.
- The hotter regions are then **LESS DENSE** than colder regions ... ... and therefore **RISE UP** through them, and start to cool.
- The colder regions now replace the hotter regions and ... ... this constant cycle is called a **CONVECTION CURRENT.**

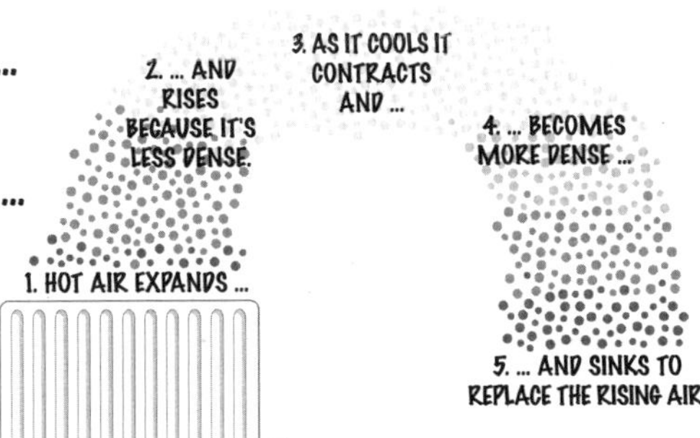

OCEAN CURRENTS

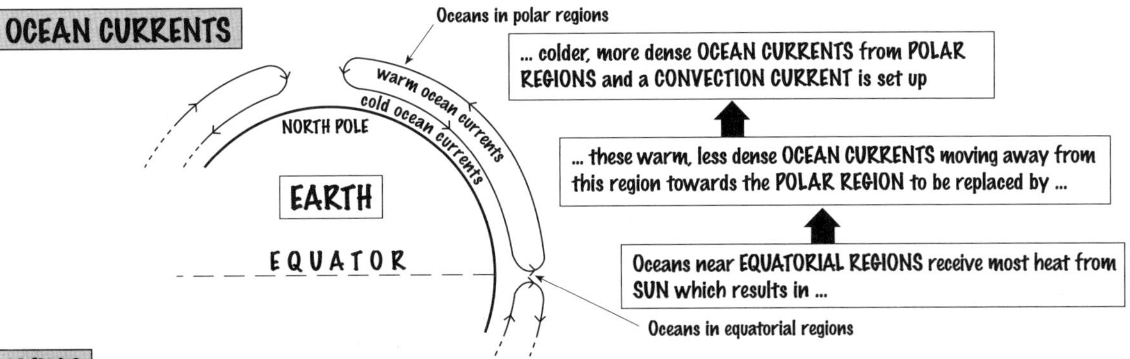

WIND

- Certain parts of the EARTH become warmer than other parts due to the **HEAT ENERGY FROM THE SUN** ...
- ... causing AIR above the warm parts to RISE, resulting in COOLER AIR from the colder parts ...
- ... moving in to take its place.
- This is WIND and causes LAND and SEA BREEZES - the classic examples!

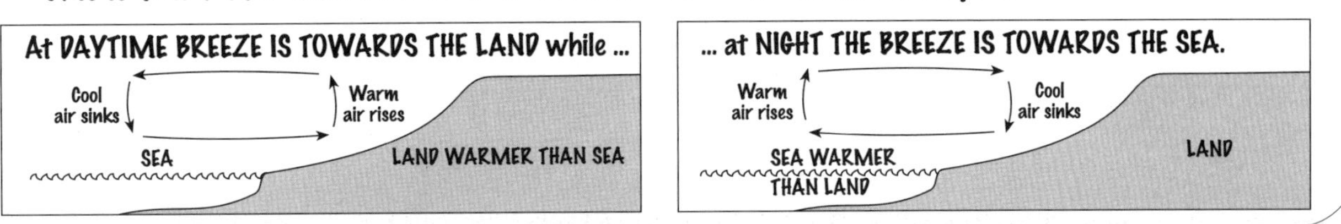

KEY POINTS:

- Heat is transferred through a solid by conduction.
- Heat is transferred through liquids and gases by convection.

3. RADIATION

- This is the transfer of energy by WAVES.
- Hot objects emit infra-red radiation which can pass through a **VACUUM** and be **REFLECTED**.
- The hotter the object the more heat energy it radiates.
- The amount of radiation **EMITTED** by an object also depends on its **SURFACE** ...

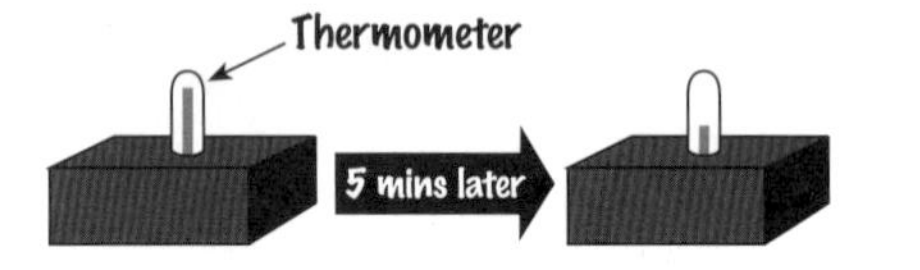

DARK MATT SURFACES EMIT MORE RADIATION THAN LIGHT SHINY SURFACES AT THE SAME TEMPERATURE

- ... while the amount of radiation **ABSORBED** by an object also depends on its **SURFACE**.

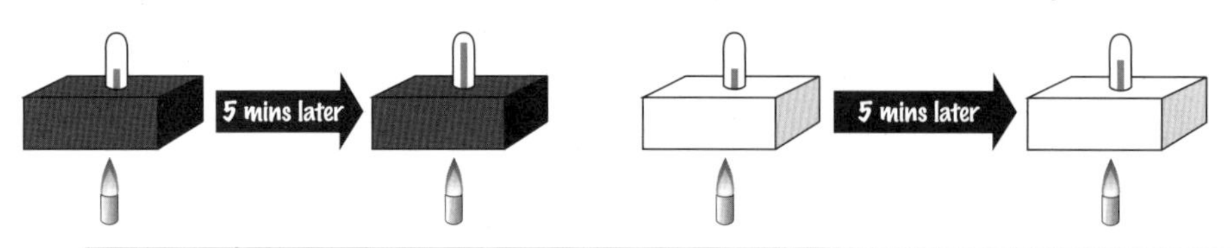

DARK MATT SURFACES ARE BETTER ABSORBERS (POORER REFLECTORS) OF RADIATION THAN LIGHT SHINY SURFACES

CARBON DIOXIDE ABSORBS I-R RADIATION
... SO HEAT RADIATED BACK INTO SPACE BY EARTH IS ABSORBED BY CO_2 IN ATMOSPHERE.
IT IS THEN RE-RADIATED BACK TOWARDS THE EARTH.
THIS IS THE GREENHOUSE EFFECT AND IT'S EFFECT IS WORSENING WITH INCREASING CO_2.

4. EVAPORATION

RATE OF EVAPORATION OF WATER DEPENDS ON ...

1. SURFACE AREA OF WATER
Increase in surface area ... increases the evaporation ... as more water molecules are near the surface.

2. HUMIDITY OF AIR
Increase in the humidity decreases the evaporation ... as air above water already saturated with water molecules.

3. MOVEMENT OF AIR
Increased movement of air ... increases evaporation ... as any water molecules in air are 'blown away'.

4. TEMPERATURE
Increased temp. of water ... increases evaporation ... as more molecules have sufficient energy to escape.

- Liquid particles which have more energy than normal ...
- ... are able to escape from the surface of the liquid ...
- ... because they can overcome the forces of attraction ...
- ... of the less energetic liquid particles left behind.

KEY POINTS:

- Radiation is the transfer of energy from hot objects.
- Transfer of heat due to loss of particles from the surface of a liquid is evaporation.

REDUCING HEAT LOSSES

It is very important that any UNWANTED energy transfers from a HOT OBJECT to a COLDER OBJECT

... are kept to a MINIMUM as very simply ...

... ENERGY COSTS MONEY!!

One material which forms the basis of most energy saving devices is **AIR** !!!

1. THE HOUSE

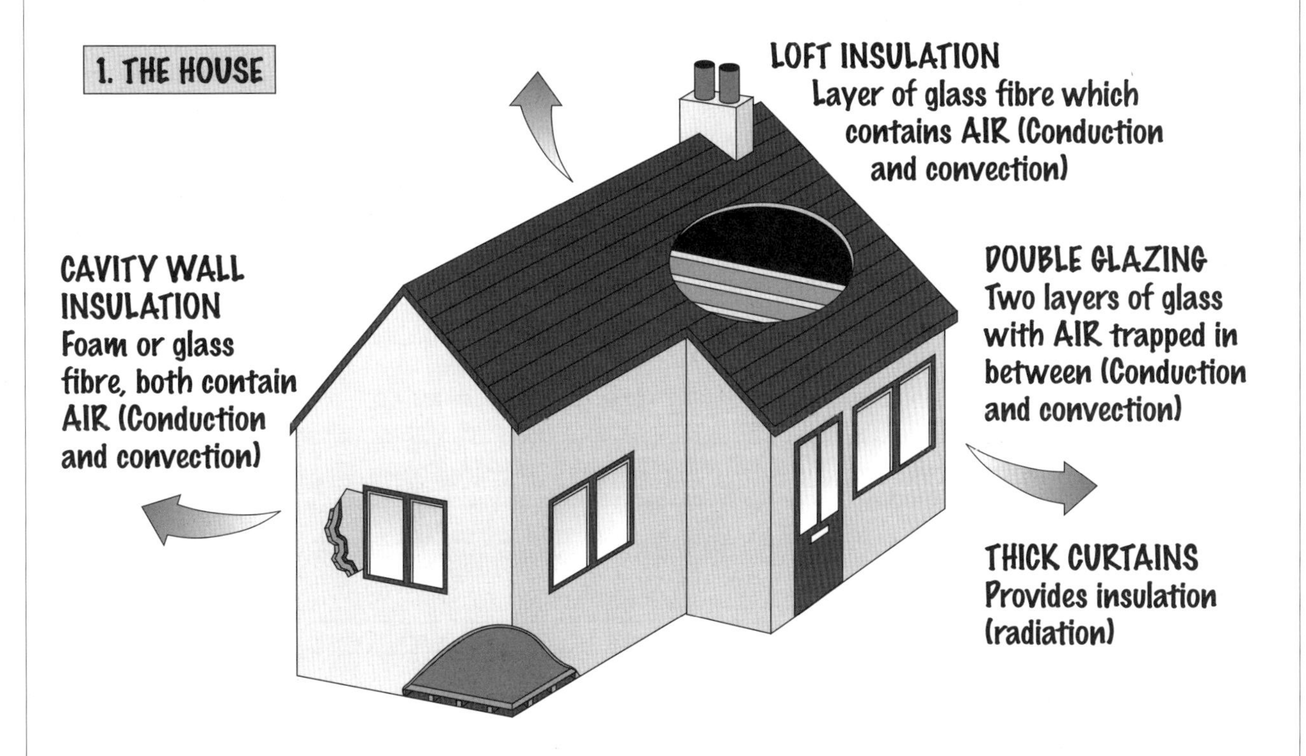

2. VACUUM FLASK

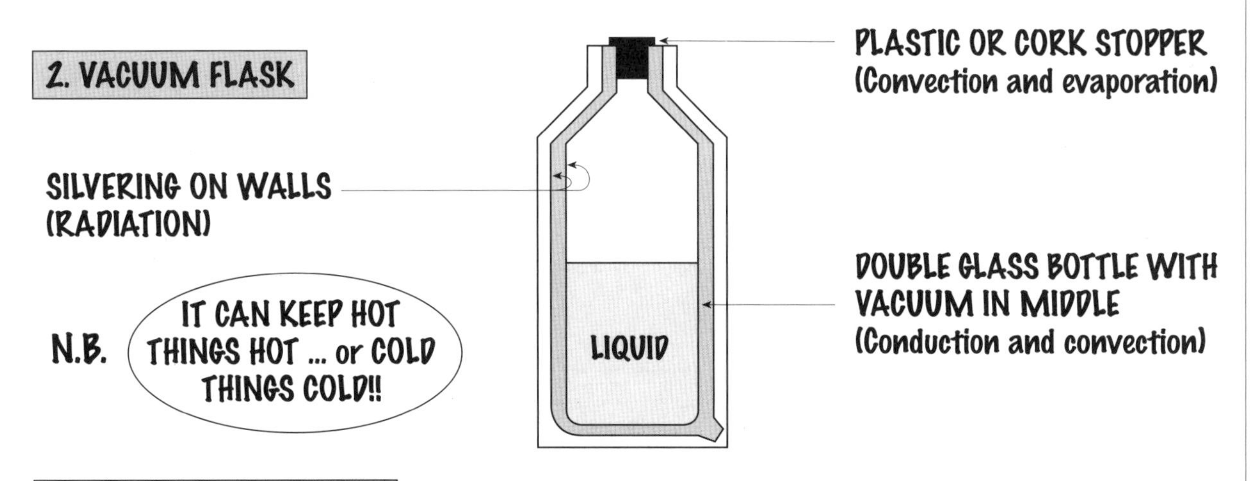

3. CLOTHING AND BEDDING

- Both made from materials that are themselves GOOD INSULATORS, however ...
- ... the insulating properties of each one is INCREASED ...
- ... because both also **TRAP AIR** within the material.

KEY POINTS:

- Air is a material which forms the basis of most energy saving devices.

RADIOACTIVITY I

RADIOACTIVITY...

- ... is the **SPONTANEOUS** (i.e. naturally occurring) **EMISSION** of **ENERGY** from atomic nuclei ...
- ... as a result of the **BREAKDOWN** of **UNSTABLE NUCLEI** and it is ...
- ... a random process which is **UNAFFECTED** by any factors, **PHYSICAL** or **CHEMICAL**!

RADIATION can be detected by ...

1. GEIGER-MULLER TUBE

Argon gas · Wire · RADIATION · TO COUNTER · Tube · Pulse of current

- When RADIATION enters TUBE it creates IONS between WIRE and TUBE.
- EFFECT is just like a 'PULSE of current which is registered by a COUNTER ...
- ... and so it's possible to MEASURE the AMOUNT of radiation which enters.

(YOU DO NOT NEED TO KNOW HOW A G-M TUBE WORKS)

2. PHOTOGRAPHIC FILM

Sheet of lead with hole in middle · RADIATION · Photographic film

- PHOTOGRAPHIC FILM is 'BLACKENED' by RADIATION
- the MORE it is EXPOSED ...
- ... the 'BLACKER' the film.

BACKGROUND RADIATION

This is **RADIATION THAT OCCURS ALL AROUND US.**

The **SOURCE** of this radiation may be **NATURAL** or **MAN-MADE** (due to human activity).

NATURAL

1. ROCKS (e.g. granite) either below or above surface of Earth contain naturally occurring atoms which are radioactive. Decay may produce RADON GAS, also radioactive, which may seep into houses and be breathed in.
2. COSMIC RADIATION from outer space.

MAN-MADE

1. MEDICAL, from the use of X rays mainly.
2. NUCLEAR INDUSTRY e.g. nuclear power stations including its waste material and fallout from weapon testing.

RADIOACTIVE EMISSIONS - types of and properties

There are THREE TYPES:	1. ALPHA (α) PARTICLES	2. BETA (β) PARTICLES	3. GAMMA (γ) RADIATION (part of the E-M spectrum)
PENETRATING POWER:	LOW	MODERATE	HIGH
BLOCKED BY:	PAPER	THIN SHEETS OF ALUMINIUM	CONCRETE/LEAD

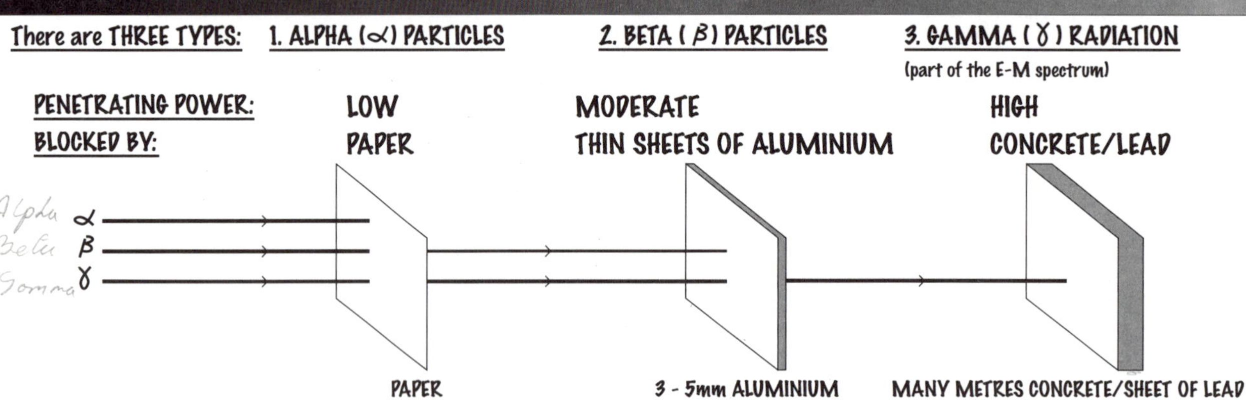

THEY ARE:	HELIUM NUCLEI (^4_2He) i.e. TWO PROTONS + TWO NEUTRONS	HIGH ENERGY ELECTRONS ($^0_{-1}e$) i.e. FAST MOVING	ELECTROMAGNETIC RADIATION WITH VERY HIGH FREQUENCY

RADIATION is EMITTED from UNSTABLE NUCLEI of atoms called RADIO-ISOTOPES ...
... ISOTOPES are ATOMS which have the SAME No. OF PROTONS BUT DIFFERENT No. OF NEUTRONS than the STABLE ATOM.

KEY POINTS:

- Alpha particles, Beta particles and Gamma radiation are the three types of radioactive emissions.
- Radiation that occurs all around us is background radiation.

EFFECTS AND USES OF RADIOACTIVITY

THE GOOD

1. MEDICINE

- To kill CANCEROUS CELLS ...
- ... use a γ source and a CALCULATED DOSE.
- STERILISING MEDICAL INSTRUMENTS ... kills all germs.
- As a TRACER inside body ...
- ... suitable SOURCE ...
- ... and HALF LIFE essential!!

2. INDUSTRY

- THICKNESS CONTROL for making PAPER and ALUMINIUM FOIL ...
- ... as amount of ABSORPTION depends on THICKNESS.
- As a TRACER to detect LEAKS in PIPES ...
- ... UPTAKE of RADIOACTIVE FERTILISERS by plants ...
- ... suitable SOURCE and HALF LIFE essential!!

THE BAD

- α, β and γ radiation can cause ...
- ... DAMAGE to ...
- ... LIVING CELLS in LIVING ORGANISMS ...

... AND THE UGLY!

- ... which can cause in ORGANS ...
- ... CANCER including LEUKAEMIA (cancer of the blood) and ...
- ... STERILITY or ABNORMALITIES IN CHILDREN BORN ...

An increase in the use of radioactive materials does produce ... SOCIAL ... ECONOMIC ... and ENVIRONMENTAL problems!

HALF LIFE

This is the TIME it takes for:

- HALF A GIVEN NUMBER OF RADIOACTIVE ATOMS (PARENT ATOMS) ...
- ... TO DECAY TO DIFFERENT ATOMS.

where ○ = PARENT ATOM.
● = NEW ATOM FORMED AFTER PARENT ATOM HAS DECAYED.

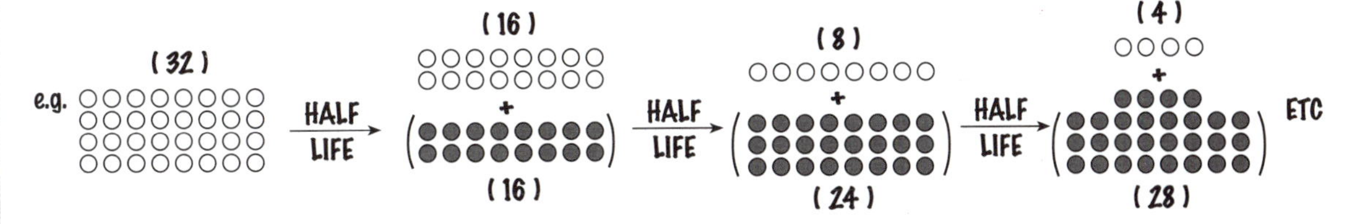

However this definition of HALF-LIFE is very IMPRACTICAL and for calculations the following alternative definition is far more useful. The HALF-LIFE is the TIME taken for ...

- THE COUNT RATE (NUMBER OF PARENT ATOMS WHICH ARE DECAYING IN A CERTAIN TIME, MEASURED USING A G-M TUBE) ... TO FALL TO HALF ITS INITIAL VALUE.

KEY POINTS:

- Radioactivity has many beneficial uses, however it can also result in cells being damaged which can lead to cancer.
- The time taken for half a given number of radioactive atoms to decay to different atoms is called the half-life.

CALCULATION OF HALF-LIFE

EXAMPLE The table below shows the results of count rate against time for a radioactive isotope. From the information calculate the **HALF-LIFE** of the isotope.

TIME (minutes)	0	5	10	15	20	25	30
COUNT RATE (counts/minute)	200	160	124	100	80	62	50

There is a **QUICK** and a **NOT SO QUICK** way of calculating the **HALF-LIFE** ... however **BOTH WAYS ARE EASY!!**

The QUICK and EASY WAY

Remember ...

... HALF-LIFE = TIME FOR COUNT RATE TO HALVE.

From our table ...

COUNT RATE	TIME	HALF-LIFE
200 → 100	0 → 15	15 mins
160 → 80	5 → 20	15 mins
124 → 62	10 → 25	15 mins
	Average =	15 mins

The NOT SO QUICK but still EASY WAY!!

You will be asked to draw a graph of COUNT RATE against TIME.

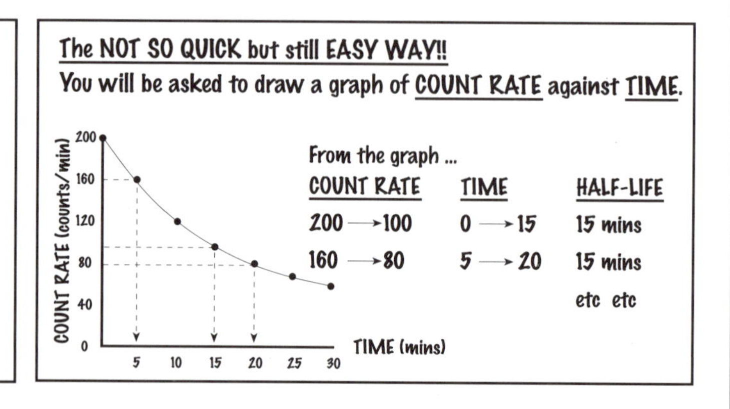

NB IF YOU ARE ALSO GIVEN THE BACKGROUND RADIATION COUNT RATE, YOU MUST TAKE THIS READING AWAY FROM EACH COUNT RATE TO GIVE THE CORRECTED COUNT RATE BEFORE YOU START TO WORK OUT THE HALF-LIFE.

DATING OF MATERIALS

- If the **HALF-LIFE** of a radioisotope is known then since certain materials contain **RADIOISOTOPES** which **DECAY** to produce **STABLE ISOTOPES** (which don't decay), ... it is possible to **DATE THE MATERIAL** ...
- ... **IGNEOUS ROCKS** contain **URANIUM -238** which decays to **LEAD** (U-238 has a half-life of 4,500,000,000 years!!!) and ...
- ... **WOOD** and **BONES** contain **CARBON-14** which decays when the organism dies!!

EXAMPLE

A very small sample of dead wood has a count rate of 1000 over a period of time, whereas the same mass of 'live' wood has a count rate of 4000 over an identical time period. If the half life of C-14 is 6000 years, calculate the age of the wood.

WE NEED TO ASSUME THAT THE DEAD WOOD, WHEN ALIVE, WOULD ALSO HAVE HAD A COUNT RATE OF 4000 DUE TO C-14.

ORIGINAL COUNT RATE **4000** —HALF-LIFE→ **2000** —HALF-LIFE→ **1000** PRESENT COUNT RATE

Therefore the C-14 has taken 2 x HALF LIVES TO DECAY TO ITS CURRENT COUNT RATE.

Therefore age of wood = 2 x 6000

= 12000 years

KEY POINTS:

- A count rate against time graph can be used to calculate the half-life of a radioactive isotope.
- It is possible to date certain materials since they contain radioisotopes which decay.

THE SOLAR SYSTEM AND THE UNIVERSE I — E 12

All BODIES including the SUN, EARTH, MOON and other PLANETS attract each other with a FORCE called GRAVITY. This GRAVITATIONAL FORCE between two bodies is the main factor which controls the movement of a small body around a larger one.

THE PLANETS
- The planets (including Earth) are NON LUMINOUS BODIES ...
  ... which orbit STARS (in our case the SUN).
- We see the OTHER PLANETS because ...
  ... LIGHT from the SUN, REFLECTS off them.
- The ORBITS of planets are ELLIPTICAL (slightly squashed circles) ...
  ... in the SAME PLANE (except Pluto) with the SUN at the centre.

THE SUN
- The SUN is a STAR ...
  ... and like all stars ...
  ... is a SOURCE of LIGHT ...
  ... and other forms of ...
  ... ELECTROMAGNETIC RADIATION.

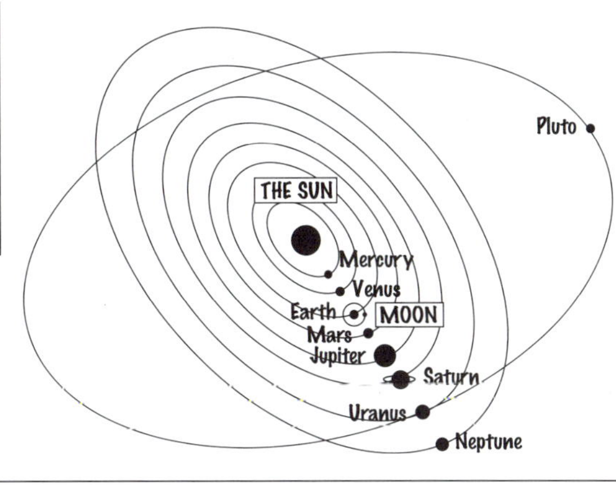

MOONS
- These are BODIES that
  ... ORBIT the PLANETS.
- Not all planets have moons ...
  ... but the EARTH has ONE ...
  ... called the MOON!!

N.B.
YOU NEED TO KNOW THE ORDER OF THE PLANETS FROM THE SUN OUTWARDS.

THE PLANETS ARE NATURAL SATELLITES OF THE SUN WHILE MOONS ARE NATURAL SATELLITES OF THE PLANETS.

THE EARTH

SUN LIGHT N AXIS DARK S

- Spins on a TILTED AXIS ...
  ... completing ONE REVOLUTION EVERY 24 HOURS, giving us DAY and NIGHT ...
  ... which explains the APPARENT MOVEMENT OF THE STARS AT NIGHT ...
  ... REMEMBER, it isn't THE STARS THAT ARE MOVING AROUND BUT THE EARTH, SPINNING ON ITS AXIS!
- Completes ONE ORBIT of the SUN every 365 ¼ days (an EARTH YEAR) giving us the DIFFERENT SEASONS.

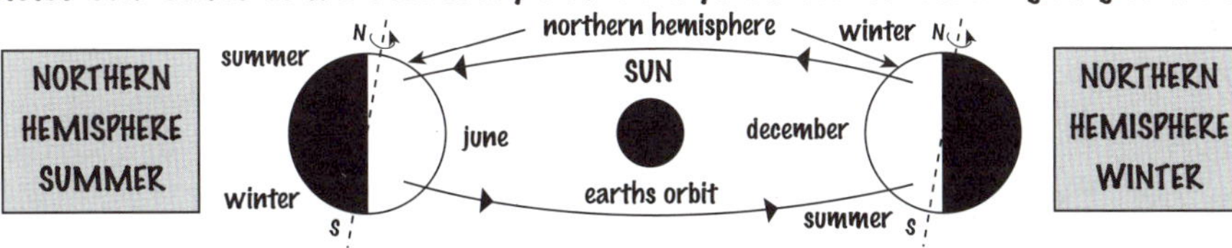

- Northern Hemisphere tilted TOWARDS THE SUN ...
- ... so DAYTIME IS LONGER THAN NIGHT TIME ...
- ... and SUN RISES HIGHER IN THE SKY ...
- ... delivering MORE ENERGY to this hemisphere ...
- ... which makes it WARMER (hopefully!)

- Northern Hemisphere tilted AWAY FROM SUN ...
- ... so NIGHT TIME IS LONGER THAN DAY TIME ...
- ... and SUN IS LOWER IN SKY ...
- ... and so delivers LESS ENERGY to this hemisphere ...
- ... which makes it COLDER.

GRAVITATIONAL FORCE BETWEEN TWO BODIES - inverse square relationship

Consider TWO BODIES a DISTANCE 'D' apart and the GRAVITATIONAL FORCE between them is 'F' then ...

- ... if their distance apart becomes '2D' (TWICE ORIGINAL) ...
  - ... the gravitational force between the two bodies becomes $\frac{'F'}{4}$ (ONE QUARTER ORIGINAL!) and ...
- ... if their distance apart becomes '3D' (THREE TIMES ORIGINAL)
  - ... then gravitational force between the two bodies becomes $\frac{'F'}{9}$ (ONE NINTH ORIGINAL!)

and so on ...

KEY POINTS:

- The planets in order from the Sun are:
  Mercury, Venus, Earth, Mars, Jupiter, Saturn, Uranus, Neptune and Pluto.
- The gravitational force between two bodies obeys the inverse square relationship.

ARTIFICIAL SATELLITES - in orbit around the Earth

They have many uses:

1. OBSERVATION OF THE EARTH

- Military purposes for 'spying'... capable of taking pictures of minute detail.
- Take photographs of 'disasters' ... floods, earthquakes, crop failures.

2. WEATHER MONITORING

- Weather satellites have a ... low polar orbit over North and South pole.
- They collect information about the atmosphere ... including cloud photographs and monitor their ... ... movement so weather forecasts can be made.

3. EXPLORATION OF THE SOLAR SYSTEM

- Space telescope (e.g. Hubble) ... in orbit above the Earth's atmosphere.
- Solar system and beyond observed with no interference as atmosphere absorbs and scatters light ... ... while clouds and weather storms ... also interfere with the light.

4. COMMUNICATIONS SYSTEMS

- • Radio, TV, Telephone links ... for places far apart.
- This is a geosynchronous satellite ... as it moves at exactly the same rate ... as the Earth revolves ... ... (takes 24 hrs to go around).
- • Remains at same position when viewed from earth.

ASTEROIDS AND COMETS

ASTEROIDS are a band of rock debris which occupy a belt between the orbits of MARS and JUPITER while ...

COMETS ...

- have a CORE OF FROZEN GAS and DUST ... ... and an ELLIPTICAL ORBIT around the sun, ... <u>WHICH IS IN A DIFFERENT PLANE FROM THAT OF THE PLANETS.</u>

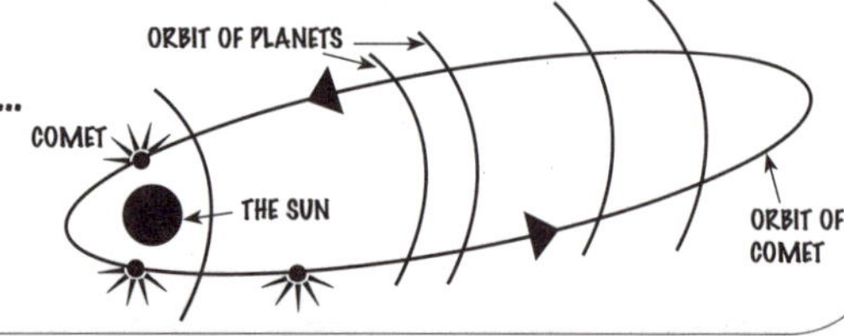

KEY POINTS:

- Artificial satellites can be used for the observation of the Earth, monitoring of the weather, exploration of the solar system and for communication.
- Asteroids are a band of rock debris and comets have a core of frozen gas and dust.

THE UNIVERSE

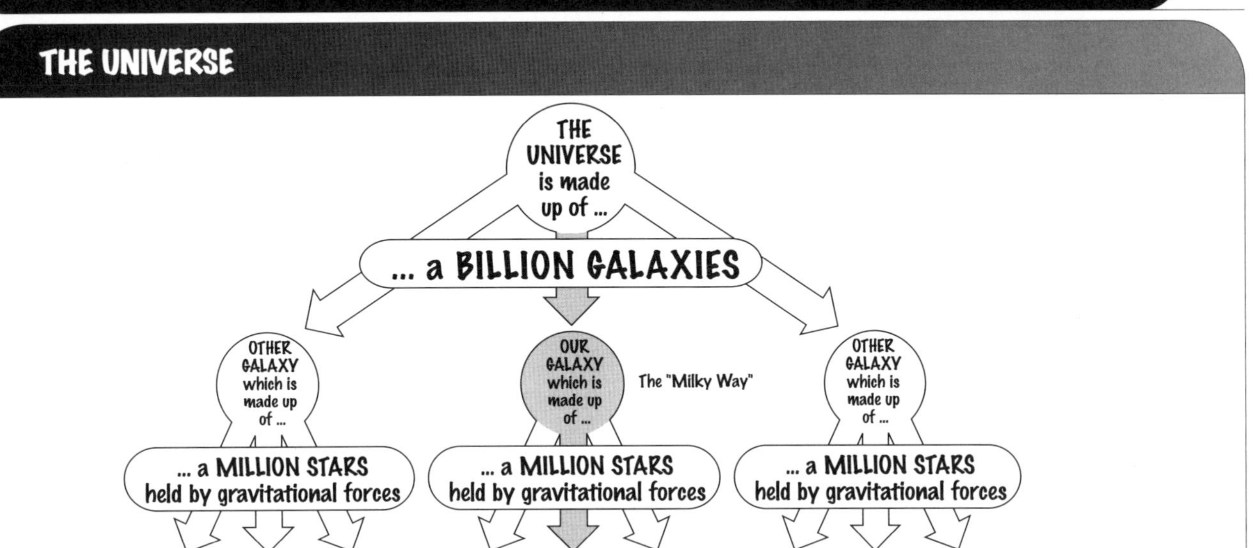

MEASURING THE UNIVERSE - LIGHT YEARS

A LIGHT YEAR is the DISTANCE travelled by LIGHT in ONE EARTH YEAR ...

- ... and is equal to 9,500 000 000 000 km!!! (light does travel at 300,000,000m/s - see FT.11).

We use this unit as the distances involved are so enormous and mind blowing!...

- Our nearest star (ALPHA CENTAURI) is 4.3 LIGHT YEARS away while ...
- ... our nearest galaxy (ANDROMEDA) is a mere 2,200,000 LIGHT YEARS away!!

ORGANISATION OF THE SOLAR SYSTEM - THE THEORIES

For over 2000 years there have been a variety of different ideas concerning the arrangement of the solar system and beyond. The acceptance or rejection of any particular idea has depended on the social and historical context in which it was developed and proposed. You are not expected to recall specific ideas.

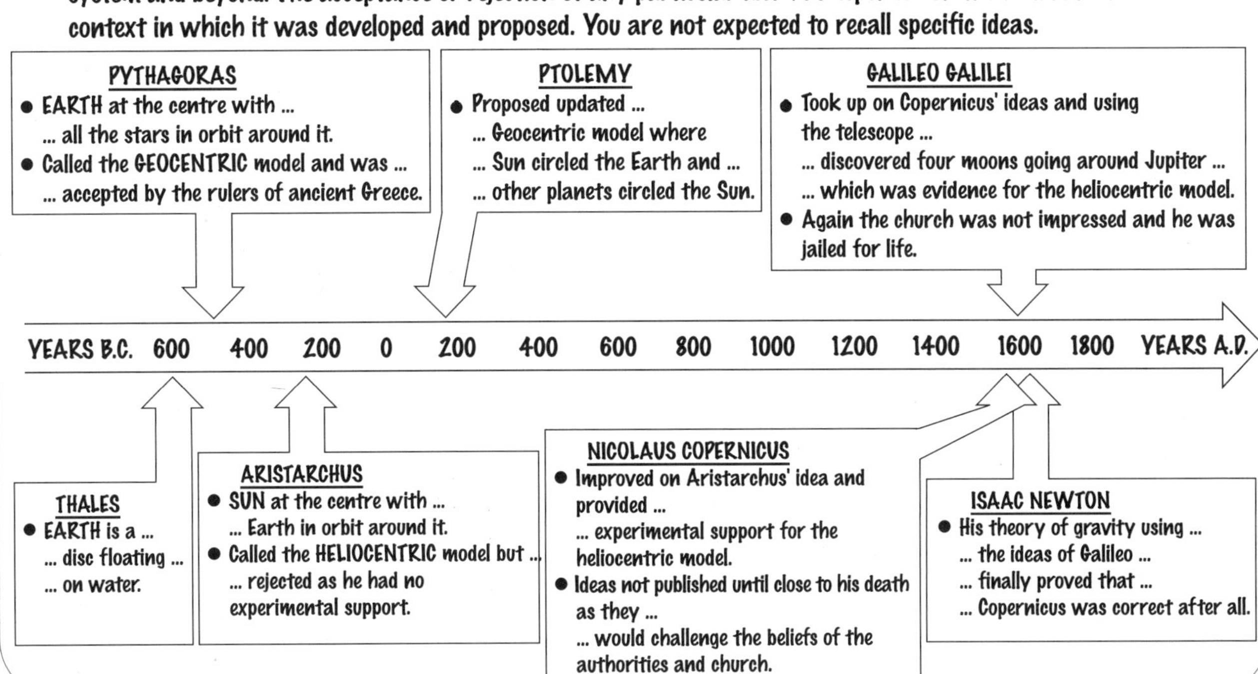

KEY POINTS:

- The Universe is made up of a billion galaxies with each galaxy containing a million stars.
- A light year is the distance travelled by light in one earth year.
- Two theories about the organisation of the solar system have been the Geocentric model and the Heliocentric model.

1. What are NON-RENEWABLE energy resources?
2. How are they used to generate electricity?
3. Why is it important to conserve these resources?
4. What are RENEWABLE energy resources?
5. Which resources are used for generating electricity?
6. Why are WOOD and FOOD called renewable energy resources?
7. Why is the SUN the ORIGINAL ENERGY SOURCE?
8. How was COAL formed?
9. What is NUCLEAR FUSION?
10. What is an EXOTHERMIC and ENDOTHERMIC REACTION?
11. What is activation energy?
12. Name TEN ENERGY FORMS.
13. Which ENERGY TRANSFERS take place when you switch a foodmixer on?
14. Write down the equation for EFFICIENCY.
15. For every 1000 Joules of chemical energy put into a car, only 240 Joules are transferred as movement energy. Calculate its EFFICIENCY.
16. Write down 4 main facts about transferring heat energy by CONDUCTION.
17. Write down 4 main facts about transferring heat energy by CONVECTION.
18. How are OCEAN CURRENTS and WIND created?
19. What is RADIATION?
20. How does a) temperature b) surface conditions and c) amount of carbon dioxide in the atmosphere affect the ABSORPTION and EMISSION of radiation?
21. What is EVAPORATION?
22. Write down as many ways as possible for preventing heat loss from a house by a) CONDUCTION and b) CONVECTION.
23. How does a VACUUM FLASK keep hot things hot or cold things cold?
24. Why are CLOTHING and BEDDING good insulators?
25. What is radioactivity?
26. How is photographic film affected by radiation?
27. What is background radiation? Name 2 examples of a) NATURAL and b) MANMADE radiation sources.
28. What are the three types of radiation?
29. Draw a diagram to show how alpha, beta and gamma can be stopped.
30. What are a) alpha particles b) beta particles c) gamma rays?
31. What are isotopes?
32. What are the a) GOOD b) BAD and c) UGLY effects of radiation?
33. Name one example of how increased use of radioactive materials has produced a) social b) economic and c) environmental problems.
34. What is half-life?
35. If a radioisotope has a half-life of 5 hours what fraction of the isotope will not have decayed after 10 hours?
36. If a radioisotope has a half-life of 2 hours what fraction of the isotope will have decayed after 10 hours?
37. The table below shows the results of count rate against time for a radioactive isotope.

Time (minutes)	0	10	20	30	40	50	60
Count rate (counts/minute)	400	330	250	200	164	122	96

    Calculate the half-life using the QUICK method.
38. Draw a graph for the results above and use it to calculate the half-life.
39. If the BACKGROUND RADIATION count rate is 10 counts/minute, work out the corrected count rate for the results above and draw another graph to show corrected count rate against time. From your graph calculate the half-life.
40. Repeat the bottom example on E.11 with a dead wood count rate of 1250 and a live one of 10,000.
41. Again repeat the same example with a dead wood count rate of 200 and a live one of 6,400.
42. What is the Sun?
43. What is the order of the planets from the Sun outwards?
44. What is the inverse square relationship for any two bodies?
45. Name any two uses for artificial satellites.
46. What are a) asteroids, b) comets?
47. What is the Universe made up of?
48. What is a light year?
49. What is the GEOCENTRIC model of the solar system?
50. What is the HELIOCENTRIC model of the solar system?

Periodic Table of the Elements

Key

Mass number →	1
	H hydrogen
Proton number (Atomic number) →	1

1	2											3	4	5	6	7	0
																	4 **He** helium 2
7 **Li** lithium 3	9 **Be** beryllium 4											11 **B** boron 5	12 **C** carbon 6	14 **N** nitrogen 7	16 **O** oxygen 8	19 **F** fluorine 9	20 **Ne** neon 10
23 **Na** sodium 11	24 **Mg** magnesium 12											27 **Al** aluminium 13	28 **Si** silicon 14	31 **P** phosphorus 15	32 **S** sulphur 16	35 **Cl** chlorine 17	40 **Ar** argon 18
39 **K** potassium 19	40 **Ca** calcium 20	45 **Sc** scandium 21	48 **Ti** titanium 22	51 **V** vanadium 23	52 **Cr** chromium 24	55 **Mn** manganese 25	56 **Fe** iron 26	59 **Co** cobalt 27	59 **Ni** nickel 28	63 **Cu** copper 29	64 **Zn** zinc 30	70 **Ga** gallium 31	73 **Ge** germanium 32	75 **As** arsenic 33	79 **Se** selenium 34	80 **Br** bromine 35	84 **Kr** krypton 36
85 **Rb** rubidium 37	88 **Sr** strontium 38	89 **Y** yttrium 39	91 **Zr** zirconium 40	93 **Nb** niobium 41	96 **Mo** molybdenum 42	98 **Tc** technetium 43	101 **Ru** ruthenium 44	103 **Rh** rhodium 45	106 **Pd** palladium 46	108 **Ag** silver 47	112 **Cd** cadmium 48	115 **In** indium 49	119 **Sn** tin 50	122 **Sb** antimony 51	128 **Te** tellurium 52	127 **I** iodine 53	131 **Xe** xenon 54
133 **Cs** caesium 55	137 **Ba** barium 56	139 **La** lanthanum 57	178 **Hf** hafnium 72	181 **Ta** tantalum 73	184 **W** tungsten 74	186 **Re** rhenium 75	190 **Os** osmium 76	192 **Ir** iridium 77	195 **Pt** platinum 78	197 **Au** gold 79	201 **Hg** mercury 80	204 **Tl** thallium 81	207 **Pb** lead 82	209 **Bi** bismuth 83	210 **Po** polonium 84	210 **At** astatine 85	222 **Rn** radon 86
223 **Fr** francium 87	226 **Ra** radium 88	227 **Ac** actinium 89															

140 **Ce** cerium 58	141 **Pr** praseodymium 59	144 **Nd** neodymium 60	147 **Pm** promethium 61	150 **Sm** samarium 62	152 **Eu** europium 63	157 **Gd** gadolinium 64	159 **Tb** terbium 65	162 **Dy** dysprosium 66	165 **Ho** holmium 67	167 **Er** erbium 68	169 **Tm** thulium 69	173 **Yb** ytterbium 70	175 **Lu** lutetium 71
232 **Th** thorium 90	231 **Pa** protactinium 91	238 **U** uranium 92	237 **Np** neptunium 93	242 **Pu** plutonium 94	243 **Am** americium 95	247 **Cm** curium 96	247 **Bk** berkelium 97	251 **Cf** californium 98	254 **Es** einsteinium 99	253 **Fm** fermium 100	256 **Md** mendelevium 101	254 **No** nobelium 102	257 **Lw** lawrencium 103

→ The lines of elements going across are called **periods**.

↓ The columns of elements going down are called **groups**.

INDEX

PROGRESS AND REVISION CHART

PAGE No.	SECTION No.	CONTENT HEADING	SYLLAB. REF.	COVERED IN CLASS	REVISED	REVISED
5	MoL1	Life Processes And Cells	(1.1)			
6	MoL2	Blood And Nutrition I - Circulation	(1.2)			
7	Mol3	Blood And Nutrition II - Digestion	(1.2)			
8	MoL4	Blood And Nutrition III - Teeth And Enzyme Summary	(1.2)			
9	MoL5	Nervous System I - Organisation	(1.3)			
10	MoL6	Nervous System II - Reflex Action	(1.3)			
11	MoL7	Nervous System III - The Eye	(1.3)			
12	MoL8	Hormones	(1.4)			
13	MoL9	Homeostasis I - The Excretory System	(1.5)			
14	MoL10	Homeostasis II - Kidney Failure And A.D.H	(1.5)			
15	MoL11	Homeostasis III - Negative Feedback Systems	(1.5)			
16	MoL12	Maintenance Of Life Summary Questions				
17	MoS1	Reproduction & Inheritance I - Meiosis & Mitosis	(2.1/2.2)			
18	MoS2	Reproduction & Inheritance II - Chromosomes, Genes & Alleles	(2.1/2.2)			
19	MoS3	Reproduction And Inheritance III - Gender Determination & Monohybrid Inheritance	(2.1)			
20	MoS4	Reproduction And Inheritance IV - Genetic Diseases	(2.1)			
21	MoS5	Reproduction And Inheritance V - Genetic Engineering. The Work Of Mendel	(2.1)			
22	MoS6	Adaptation And Competition	(2.3)			
23	MoS7	Variation And Evolution I - Variation	(2.4)			
24	MoS8	Variation And Evolution II - Mutations And Breeding	(2.4)			
25	MoS9	Variation And Evolution III - Natural Selection	(2.4)			
26	MoS10	Variation And Evolution IV - Fossil Records	(2.4)			
27	MoS11	Humans And The Environment I - Pollution	(2.5)			
28	MoS12	Humans And The Environment II - Ecosystems	(2.5)			
29	MoS13	Health	(2.6)			
30	MoS14	Maintenance Of The Species Summary Questions				
31	SC1	Atomic Structure	(3.1)			
32	SC2	The Periodic Table I - Classification Of Elements & Groups O	(3.2)			
33	SC3	The Periodic Table II - Group I	(3.2)			
34	SC4	The Periodic Table III - Group 7	(3.2)			
35	SC5	Reactivity Series I - The Reactivity Series	(3.3)			
36	SC6	Reactivity Series II - Displacement Reactions	(3.3)			
37	SC7	Reactivity Series III - Extraction And Corrosion Of Metals	(3.3)			
38	SC8	Acids And Bases	(3.4)			
39	SC8	Rates Of Reaction I	(3.5)			
40	SC9	Rates Of Reaction II	(3.5)			
41	SC10	Useful Products From Oil I - Crude Oil	(3.6)			
42	SC11	Useful Products From Oil II - Alkanes, Alkenes And Polymers	(3.6)			
43	SC12	Useful Products From Oil III - Burning Hydrocarbons	(3.6)			